The Species Problem

There is longstanding disagreement among systematists about how to divide biodiversity into species. Over twenty different species concepts are used to group organisms, according to criteria as diverse as morphological or molecular similarity, interbreeding and genealogical relationships. This, combined with the implications of evolutionary biology, raises the worry either that there is no single kind of species, or that species are not real.

This book surveys the history of thinking about species from Aristotle to modern systematics in order to understand the origin of the problem, and advocates a solution based on the idea of the division of conceptual labor, whereby species concepts function in different ways – theoretically and operationally. It also considers related topics such as individuality and the metaphysics of evolution, and how scientific terms get their meaning. This important addition to the current debate will be essential for philosophers and historians of science, and for biologists.

RICHARD A. RICHARDS is Associate Professor, Department of Philosophy, University of Alabama. He has published in major journals on a variety of topics in the philosophy of science and biology, including phylogenetic inference, theory choice, taxonomy, and species concepts. He has also written extensively on Darwin's views about artificial selection and domestic breeding, and contributed to *The Cambridge Companion to the Origins of Species* (Cambridge, 2009).

CAMBRIDGE STUDIES IN
PHILOSOPHY AND BIOLOGY

General Editor
Michael Ruse, *Florida State University*

Advisory Board
Michael Donoghue, *Yale University*
Jean Gayon, *University of Paris*
Jonathan Hodge, *University of Leeds*
Jane Maienschein, *Arizona State University*
Jesús Mosterín, *Instituto de Filosofía (Spanish Research Council)*
Elliott Sober, *University of Wisconsin*

Recent Titles
Alfred I. Tauber *The Immune Self: Theory or Metaphor?*
Elliott Sober *From a Biological Point of View*
Robert Brandon *Concepts and Methods in Evolutionary Biology*
Peter Godfrey-Smith *Complexity and the Function of Mind in Nature*
William A. Rottschaefer *The Biology and Psychology of Moral Agency*
Sahotra Sarkar *Genetics and Reductionism*
Jean Gayon *Darwinism's Struggle for Survival*
Jane Maienschein and Michael Ruse (eds.) *Biology
and the Foundation of Ethics*
Jack Wilson *Biological Individuality*
Richard Creath and Jane Maienschein (eds.) *Biology and Epistemology*
Alexander Rosenberg *Darwinism in Philosophy,
Social Science, and Policy*
Peter Beurton, Raphael Falk and Hans-Jörg Rheinberger (eds.) *The
Concept of the Gene in Development and Evolution*
David Hull *Science and Selection*
James G. Lennox *Aristotle's Philosophy of Biology*
Marc Ereshefsky *The Poverty of the Linnaean Hierarchy*
Kim Sterelny *The Evolution of Agency and Other Essays*
William S. Cooper *The Evolution of Reason*
Peter McLaughlin *What Functions Explain*

The Species Problem: A Philosophical Analysis

RICHARD A. RICHARDS

University of Alabama

CAMBRIDGE
UNIVERSITY PRESS

CAMBRIDGE
UNIVERSITY PRESS

University Printing House, Cambridge CB2 8BS, United Kingdom

Cambridge University Press is part of the University of Cambridge.

It furthers the University's mission by disseminating knowledge in the pursuit of education, learning and research at the highest international levels of excellence.

www.cambridge.org
Information on this title: www.cambridge.org/9781107541078

© Richard A. Richards 2010

First published 2010
First paperback edition 2015

A catalogue record for this publication is available from the British Library

ISBN 978-0-521-19683-3 Hardback
ISBN 978-1-107-54107-8 Paperback

For Rita Snyder, my partner in life and dance.

Contents

Acknowledgments

My debts to Michael Ruse are many. I have benefited from his guidance and encouragement in writing this book, from his generous support in attending workshops and conferences at Florida State University, but also from the example he has set. He has shown time and again the value in combining the insights gleaned from the history of science with those from the philosophy of science. Hilary Gaskin, of the Cambridge University Press, has been very helpful and supportive. I have also benefited extensively from the insights and criticisms of Mary Winsor, Phillip Sloan and Manfred Laubichler. I am grateful as well for the insights due to my discussions with Richard Mayden, Max Hocutt and James Otteson. There are others who have been invaluable, not just for my understanding of the species problem, but in the development of the general philosophical framework in which this discussion occurs. Worthy of special acknowledgment are David Hull and Marc Ereshefsky, for their insights and efforts. I must emphasize, however, that many of those to whom I owe these debts will undoubtedly disagree with the view presented here. Finally, I will be forever grateful for the encouragement and support of my parents, Richard and Annette Richards, and for the guidance, criticism, friendship and support of my mentor, Peter Achinstein.

1

The species problem

Species are kinds of living things. This way of thinking about life seems to go back at least to Plato and Aristotle, who used the term *eidos* that meant in one sense, the appearance or form of a thing. For the naturalists who came after, species were also the relatively distinct groupings of individual organisms that were more or less similar in appearance and behavior, and that sometimes interbred. Since the development of a hierarchical taxonomy by Linnaeus in the eighteenth century, those organisms that had been grouped into species taxa were then grouped into more inclusive categories – genera, orders, classes, and ultimately kingdoms. Because species are the most basic groupings of organisms in this hierarchy, they are now usually regarded as the fundamental units of biodiversity. But for contemporary biologists steeped in evolutionary theory, species are much more. Darwinian evolution tells us that species are the things that are "born" in speciation from other species, change over time, produce new species, and ultimately "die" in extinction. Species are therefore also the fundamental units of evolution.

The idea of species has played a similarly significant role in philosophy. Philosophers have followed the tradition of Plato and Aristotle, as they understand it, and have treated those groups of organisms we identify as species as *natural kinds* with *essences*. In doing so, they have treated biological species as equivalent to chemical elements such as hydrogen and oxygen, and molecular kinds, such as water, that are made up of these elements. Biological species have in this way fit into a philosophical way of thinking known as *metaphysics*, which studies the basic, fundamental things and processes that exist. On this traditional essentialist approach, biological species, like hydrogen, oxygen and water, are the fundamental

1

and eternal kinds of things we find in nature. And when we divide nature into species, elements and molecular kinds, we are, in the oft-paraphrased words of Plato, "cutting nature at its joints." But Darwinian evolution has seemed by many to challenge this idea that species are natural kinds with essences. Most obviously, evolution implies that species can no longer be regarded as eternal and unchanging. If so, then how do *species* fit into our philosophical understanding of the world?

The philosophical significance of this idea of *species* extends into our understanding of human nature. In the essentialist tradition, humans have a nature because they belong to the species-kind *human*. In more modern terms, humans are the way they are by virtue of being members of the species *Homo sapiens*. The idea of human nature is therefore dependent on our ideas of what it means to be a member of a species. In the past this might have meant an understanding based on the essence of being human. But with the Darwinian challenge to this traditional picture, we also get a challenge to traditional ways of thinking about human nature. If evolution forces us to rethink the nature of species, perhaps we must also rethink the nature of human nature.

SPECIES GROUPINGS

The biological and philosophical significance we place on this idea of species is particularly striking given the difficulties we have in consistently placing organisms into species in *microtaxonomy*. Here, the main tasks are first, dividing and grouping of organisms into species; second, providing criteria for species membership. On both tasks there is substantial and pervasive disagreement among biological systematists. Given any single group of individual organisms, systematists will often disagree about the number of species represented and the criteria used in making that determination. Some of these disparities in counts are highly striking. Counts of lichen species worldwide, for instance, range from around 13,000 to 30,000 (Purvis 1997: 111). Researchers have also counted from one to ten species in the fish genus *Metriaclima*, 101 to 240 species in Mexican birds, and 9000 versus 20,000 bird species worldwide. Jody Hey cites three reasons for these disparities in species counts: count creep, lumper/splitter tendencies, and the use of different species definitions:

> Consider the case of *Metriaclima*, a genus of 10 species of fish in Lake Malawi, Africa, that was devised to replace a single species

2

Pseudotropheus zebra, on the basis of additional collections ... The first revision was a case of count creep, pure and simple – a closer look with more samples begat more species. But then another look, by others who were using the very same data ... led to the conclusion that the new genus actually contains only two species ... Nor are lumper/splitter debates limited to obscure organisms that are difficult to collect. Consider birds, which are probably the most observationally accessible animals on the planet ... Conventional classifications place the number of species world-wide at around 9,000. But some ornithologists feel that the correct count, based on a proper reevaluation of all existing collections, would end up being closer to 20,000 ... In fact, a count of endemic Mexican bird species went so far as to employ two different definitions of species; one returned a count of 101 species while the other returned a count of 249 species. (Hey 2001: 20)

Disparities due to count creep result from the fact that, when we look at more specimens, we simply see more differences and tend to postulate more species. Disparities due to lumper/splitter tendencies are a consequence of the subjective tendencies of individual researchers: some systematists are simply more prone to split groups of organisms into more species taxa than are other systematists.

But often disagreements about species counts are due to the fact that different researchers use different ways of defining and conceiving species. One researcher might, for instance, use morphological or genetic similarity to group into species, while another might use interbreeding, and yet another might appeal to history or phylogeny. In other words, one person might use a species concept based on morphological or genetic similarity, while another might use a concept based on interbreeding or phylogeny. The differences in species counts due to the use of different concepts are often striking. The turn to a *phylogenetic species concept*, for instance, has multiplied fifteen amphibian species into 140 (MacLauren and Sterelney 2008: 28). A recent survey of taxonomic research quantifies the effects of a shift to this particular species concept from other concepts, finding a 300% increase in fungus species, a 259% increase in lichen species, a 146% increase in plant species, a 137% increase among reptile species, an 88% increase among bird species, an 87% increase among mammals, and a 77% increase among arthropods. Running counter to this trend, however, there was a 50% decrease in mollusc species (Agapow et al. 2004:168). Overall, there was an increase of 48.7% when a *phylogenetic species concept* replaced other concepts. (Agapow et al. 2004:164).

THE SPECIES PROBLEM

This use of different species concepts is more troubling than the other sources of disparity in species counts. Count creep and lumper/splitter differences are surely significant problems in the analysis of biodiversity, but they do not challenge two basic assumptions behind the view that species are the fundamental units of biodiversity and evolution: first, species are real; second, there is a single kind of species thing. If species really exist and there is some single kind of species thing, then we can potentially resolve the disagreements that arise from both the observation of new specimens and different tendencies to split or group. We can in principle, for instance, identify what makes a new specimen a genuine instance of a new species, whether through genetic analysis, observation of interbreeding or some other criterion. And we can establish that some researchers, whether splitter or lumper, really are getting the classification more right than others, by reference to whatever factors are important, be they morphological, genetic, or reproductive. But if the differences in grouping are due to the use of conflicting species concepts, then it is hard to see how we can come to agree on species groupings just on the basis of more information about biodiversity and evolution. If we are using different species concepts and criteria for what counts as a species, new information is unlikely to result in agreement because we disagree about what is even relevant! Someone who uses a reproductive criterion will not treat newly discovered similarities and differences as relevant, whereas they will be relevant to someone who uses a morphological concept.

What has happened recently reinforces this pessimism. The more we learn about biodiversity and all its complexity, the worse the problem seems to become. Instead of resolving differences in the use of species concepts, new information seems to have resulted in the multiplication of species concepts. On at least one count, there are now over twenty species concepts in circulation based on morphological or genotypic similarity, mate recognition systems, ecological niche, phylogenetic history and more (Mayden 1997). This may come as a surprise to those of us who learned the *biological species concept* in our introductory biology classes, that species are groups of interbreeding or potentially interbreeding organisms. It takes only a moment to realize, however, that this concept applies only to sexually reproducing organisms and we would need at least one other species concept for the many asexual organisms we find throughout the plant and animal kingdoms. The biological

4

species concept is clearly inadequate if we are looking at asexually reproducing organisms.

As this limitation of the biological species concept suggests, sometimes the choice of species concept seems to depend on little more than
which organisms one studies. Joel Cracraft explains:

> There has been something of a historical relationship between an adopted
> species concept and the taxonomic group being studied ... Thus, for many
> decades now, ornithologists, mammalogists, and specialists from a few
> other disciplines have generally adopted a Biological Species Concept;
> most invertebrate zoologists, on the other hand, including the vast major-
> ity of systematists, have largely been indifferent to the Biological Species
> Concept in their day-to-day work and instead have tended to apply spe-
> cies status to patterns of discrete variation. Botanists have been some-
> where in the middle, although most have not used a Biological Species
> Concept. (Cracraft 2000: 4–5)

But even among those who study the same organisms, there is disagreement about which species concept is best. Those who are committed to
a method of taxonomy known as "cladistics" tend to use different concepts than those who have adopted the more traditional "evolutionary
systematics." And even those who regard themselves as cladists find little agreement. In a recent volume, five different cladistic species concepts were proposed and developed, seemingly without any resolution
(Wheeler and Meier 2000).

The bottom line is that there is pervasive disagreement about the
nature of species; and this has led to disagreement about how we should
divide and group organisms into species. Additional observation and
research offers little promise. The more we learn the worse the conflict
seems to become. This then is *the species problem*: there are multiple,
inconsistent ways to divide biodiversity into species on the basis of multiple, conflicting species concepts, without any obvious way of resolving
the conflict. No single species concept seems adequate.

SIGNIFICANCE OF THE SPECIES PROBLEM

While the problem seems to be getting worse, worries about it are not
new. In 1957, Ernst Mayr was already lamenting its persistence:

> Few biological problems have remained as consistently challenging
> through the past two centuries as the species problem. Time after time

attempts were made to cut the Gordian knot and declare the species problem solved either by asserting dogmatically that species did not exist or
by defining, equally dogmatically, the precise characteristics of species.
Alas, these pseudosolutions were obviously unsatisfactory. One might
ask: "Why not simply ignore the species problem?" This also has been
tried, but the consequences were confusion and chaos. The species is a
biological phenomenon that cannot be ignored. Whatever else the species
might be, there is no question that it is one of the primary levels of integration in the many branches of biology, as in systematics (including that
of microorganisms), genetics, and ecology, but also in physiology and in
the study of behavior. Every living organism is a member of a species, and
the attributes of these organisms can often best be interpreted in terms of
this relationship (Mayr 1957a: iii).

As suggested here, part of the significance of the species problem is its
implications for biological practice and theory. Biologists today see similar significance. Joel Cracraft acknowledges the species problem, then
explains its significance to theory and practice:

> The primary reason for being concerned about species definitions is that
> they frequently lead us to divide nature in very different ways. If we accept
> the assumption of most systematists and evolutionists that species are real
> things in nature, and if the sets of species specified by different concepts
> do not overlap, then it is reasonable to conclude that real entities of the
> world are being confused. It becomes a fundamental scientific issue when
> one cannot even count the basic units of biological diversity. Individuating
> nature "correctly" is central to comparative biology and to teasing apart
> pattern and process, cause and effect. Thus, time-honored questions in
> evolutionary biology – from describing patterns of geographic variation
> and modes of speciation, to mapping character states or ecological change
> through time, to biogeographic analysis and the genetics of speciation,
> or to virtually any comparison one might make – will depend for their
> answer on how a biologist looks at species (Cracraft 2000: 6).

If Mayr and Cracraft are right, there is much at stake here for those
who work in the biological sciences. Work in multiple areas depends on
how species are grouped, and the principles used for grouping. The species problem still looms large in evolutionary biology.

There is practical significance as well. Many problems are generated
by our inability to group organisms unambiguously into species. First
and most obviously, the application of endangered species legislation
seems to presuppose our ability to group organisms into species on the
basis of a satisfactory grouping principle and species concept. Claridge,

Dawah and Wilson recognize this in their introduction to a recent collection of articles on species:

> The prolonged wrangle among scientists and philosophers over the nature of species has recently taken on added and wider significance. The belated recognition of the importance of biological diversity to the survival of mankind and the sustainable use of our natural resources makes it a matter of very general and urgent concern. Species are normally the units of biodiversity and conservation ... so it is important that we should know what we mean by them. One major concern has been with estimating the total number of species of living organisms that currently inhabit the earth ... In addition, many authors have attempted to determine the relative contributions of different groups of organisms to the totality of living biodiversity ... Unless we have some agreed criteria for species such discussions are of only limited value (Claridge, Dawah and Wilson 1997: 2).

The pessimism of these biologists is reinforced by the conflicting accounts of species we get in one official interpretation of the *Endangered Species Act* of 1973 (ESA) sponsored by the US Department of Interior and published by the National Academy of Science. Michael Clegg, the Chair of the *Committee on Scientific Issues in the Endangered Species Act*, tells us in the introduction that "Species are objective entities that are easily recognized. Their health and needs can be assessed and sound scientific management plans can be implemented" (National Research Council 1995: ix). We then learn that the *Endangered Species Act* seems to assume some version of the *biological species concept*.

> Species of organisms are fundamental objects of attention in all societies, and different cultures have extensive literature on the history of species concepts. The Endangered Species Act defines species to include "any subspecies of fish or wildlife or plants, and any distinct population segment of any species of vertebrate fish or wildlife which interbreeds when mature." In the act, the term *species* is used in a legal sense to refer to any of these entities. In addressing its use in the ESA, one must remember, however, that species has vernacular, legal and biological meanings (National Research Council 1995: 5).

But it should be obvious that this way of conceiving species is highly problematic. First, and paradoxically, *species* now get understood in terms of *subspecies* – which is itself not obviously definable except relative to *species*. Second, the assumption of interbreeding seems to rule out non-sexually reproducing organisms. Third, not only must we worry about biological species concepts, but also the vernacular and legal concepts. Species concepts continue to multiply now on the basis of how non-biologists in different cultures think – and how legal scholars think! In spite of these further complications, Clegg remains optimistic:

> Many societies have notions of kinds of organisms, usually organisms that are large and conspicuous or of economic importance. The term *species* can be applied to many of these kinds and can be accurate as a scientific and vernacular term, because the characteristics used to differentiate species can be the same in both cases. Largely for this reason, the question of what a species is has not been a major source of controversy in the implementation of the Endangered Species Act (National Research Council 1995: 5).

But then he goes on to recognize the difficulties in identifying subspecies – which count as species in the assumed definition above:

> Greater difficulties have arisen in deciding about populations or groups of organisms that are genetically, morphologically, or behaviorally distinct, but not distinct enough to merit the rank of species – i.e., subspecies, varieties, and distinct population segments (National Research Council 1995: 5).

Notice also that the interbreeding criterion of the species definition does not appear in this passage. Rather, it appeals to morphological, genetic and behavior distinctness. It is difficult to make sense of this account of species given what the report later has to say about the history of species concepts:

> [B]iologists with different perspectives and problems in mind have different ideas about what a species is and what role it should play in particular areas of science. Some systematic biologists have declared that there is no single unit that can be called species, and, for example, that the concept of species used in classifying mosses might be quite different from that used for classifying species of birds with respect to population and genetic structure (National Research Council 1995: 51).

The authors then ask: "Why should the term *species* be so problematic? Why, after centuries of investigations, are systematic biologists unable to simply and easily tell us which groups of organisms are species and which are not?" (National Research Council 1995: 51–52). While the authors then give an answer – "speciation is a gradual process" – it should be obvious that there is more to a satisfactory answer. As already acknowledged, different species concepts are in use. Given all these complications, Clegg's optimism that "species are objective entities that are easily recognized" seems hardly warranted. As important as it may be to preserve biodiversity, doing so is clearly more complicated than Clegg acknowledges.

In actual application of the Endangered Species Act we find just the sorts of complications we might expect. On a morphological species concept, or one involving geographic isolation, we might classify the red wolf of the southeastern US as a separate species from the wolves of eastern Canada. But on other criteria, such as potential interbreeding, we might classify them together, as is implied by science journalist Carl Zimmer in a recent article in *Scientific American*:

> Wolves in the southeastern U.S. are considered a separate species, the red wolf (*Canis rufus*). This wolf has been the subject of an enormous project to save it from extinction, with a captive breeding effort and a program to reintroduce it to the wild. But the Canadian scientists argue that the red wolf is really just an isolated southern population of *C. lycaon*. If that is true, then the government has not in fact been saving species from extinction. Thousands of animals belonging to the same species are still thriving in Canada (Zimmer 2008: 73).

To complicate things further, it appears that coyotes have in fact successfully interbred with *C. lycaon*, and both groups contain DNA of the other group. On an interbreeding concept, both groups of wolves are members of a species also containing coyotes (Zimmer 2008: 72).

The differences in species counts and application of endangered species concepts have real consequences beyond the preservation of biodiversity. The turn to the *phylogenetic species concept* that multiplied species counts so dramatically, also has a cost. The authors of the survey quoted above have estimated the cost of the proliferation of species taxa, based on the fact that increased species counts will reduce the geographic range of species, that will then make more species protected.

> Any increase in the number of endangered species requires a corresponding increase in resources and money devoted toward conserving those species. For example, it has been estimated that the complete recovery of any of the species listed by the U.S. Endangered Species Act will require about $2.76 million ... Thus, recovering all species listed currently would cost around $4.6 billion. With widespread adoption of the PSC [*phylogenetic species concept*], this already formidable amount could increase to $7.6 billion, or the entire annual budget for the administering agency (U.S. Fisheries and Wildlife Services) for the next 120 years (Agapow *et al.* 2004: 169).

And as these authors then indicate, this estimate of an additional $3 billion in cost might well be conservative.

There are other practical reasons to worry about our ways of grouping organisms into species. We might, for instance, worry about the preservation of biodiversity independent of any legislative demands. Measurements of biodiversity often employ the concept of *species richness* to measure biodiversity (MacLauren and Sterelny 2008: 3). Species richness is straightforwardly dependent on species counts (higher species counts means greater species richness), so if our species counts are problematic so will be our assessments of biodiversity. The management of food sources and natural resources also often requires we know something about particular species as Joel Cracraft argues:

> The importance of species concepts is not restricted to the seemingly arcane world of systematics and evolutionary biology. They are central to solving real-world practical problems that affect people's lives and well-being. ... Consider, for example, cases in which species concepts might have important consequences: (1) a group of nematodes that attack crops, or act as vectors for plant viruses, where failure to individuate species correctly might mean that food supplies are at risk. ... (2) a group of exotic beetles that attack timber resources, where failure to individuate species correctly might mean that their place of origin could be misidentified and thus potential biological control agents overlooked (Cracraft 2000: 6–7).

Similarly, human disease vectors may well be associated with particular species (Cracraft 2000: 7). It is well known, for instance, that malaria infects humans by contact with just a few of the species of mosquito in the genus *Anopheles*. Other species do not present a risk. If so, we may need to get our species grouping right to prevent the spread of this disease. There are many other instances where species determinations have similar practical import, but the main message here should already be clear: There is a lot at stake in getting our species groupings right and that depends on getting our species concepts right.

REALISM AND PLURALISM

Lurking behind the species problem are two philosophical worries. On the first *realism* worry, we might – given the proliferation of species concepts – doubt that species are real things in nature. Cracraft expresses this worry.

Unless species concepts are used to individuate real, discrete entities in nature, they will have little or no relevance for advancing our understanding of the structure and function of biological phenomena involving those things we call species ... If species are not considered to be discrete real entities ... it implies that evolutionary and systematic biology would be based largely on units that are fictitious, whose boundaries, if drawn, are done so arbitrarily. It would also mean that most, if not all, of the processes that we ascribe to species are concoctions of the mind and have no objective reality (Cracraft 1997: 327–328).

Because species play a theoretical role, functioning within scientific theories, they have theoretical significance. And this seems to require that they be real! If species are not real, so much the worse for all our biological theories that presuppose the reality of species. Furthermore, if species are not real, it is hard to see how we can preserve them, or why we should try. Surely we do not want to waste resources on the preservation of biological fictions.

We might also worry about *pluralism*. If species are real, is there some one kind of species thing, or are there multiple kinds of species things? In other words, *species taxa* – the grouping of organisms – might be real, without the *species category* real. If so, when we use the word *species,* we are potentially referring to different kinds of things. Perhaps vertebrate species are not the same kinds of things as bacterial or fungal species. This worry has immediate scientific implications. If part of the goal of science is to get a satisfactory account of the basic things and processes in nature, then surely part of biological systematics is getting the description of biodiversity right, and this presupposes a satisfactory way of conceiving this diversity. To do this, we need to know what species are, and whether they are all ultimately the same kinds of things. And if there are different kinds of species things, then we should accommodate that fact in our species legislation and efforts to preserve biodiversity.

There are other philosophical questions raised by the species problem. First is a "metaphysical," or more precisely an "ontological," question. If species are real and a single kind of thing, what basic, fundamental kind of things are they? Are they *natural kinds* as long supposed by philosophers? Are they then also *sets* of organisms? Or are they something else? One philosophical position that has recently generated intense debate is the claim that species are not natural kinds in the traditional sense, but they are instead *spatio-temporally restricted individuals*. If so, then species are the same fundamental kinds of things as individual organisms. This counterintuitive view has its share of supporters in biological systematics, and as we shall see, much to recommend it. Other relevant philosophical topics are related to the nature and functioning of scientific concepts. How do scientific concepts such as species concepts function in science? How do they get and change meaning? How do they get applied to nature? By a careful examination of the functioning and semantics of species concepts, perhaps we can gain insights into the functioning of scientific concepts in general.

In this book, I will examine *the species problem* and the philosophical questions it raises. I shall argue that there is a promising solution, but not where we have so far been looking. The usual approach of those who want to solve the species problem is to look toward the adequacy of particular concepts. One standard way to assess the adequacy of a species concept is on its success relative to three criteria. First is the species concept universal – does it apply to all kinds of biodiversity from vertebrates to insects, plants and bacteria? As noted already, this is a criterion where the interbreeding based *biological species concept* fails, applying only to those organisms that reproduce sexually. Second, is it applicable – can it actually be applied on the basis of what we can observe in the world around us? The *biological species concept* seems to fail here as well. How can one observe actual or potential interbreeding in populations that are separated geographically? The third criterion is theoretical significance. Does the species concept in question have the right kind of theoretic significance – fit within the relevant theoretical frameworks in the right way?

David Hull has asked how standard species concepts fare against these three criteria. He began by comparing the search for an ideal species concept with the search for the Holy Grail. His pessimism is striking:

> The temptation has always been to hope that, if we can only formulate the right definition, all our problems will be solved. Enough time has passed and enough energy expended to convince quite a few of us that no magic bullet exists for the species concept ... Any species concept, no matter which one we choose, will have some shortcoming or other. Either it is only narrowly applicable, or if applicable in theory, not in practice, and so on. One problem is that different systematists have different goals for species concepts, but even those systematists who agree in principle on what a species concept should do frequently prefer different species concepts. The trouble is that we have several criteria that we would like an ideal species concept to meet, and these criteria tend to conflict. Most importantly, if a species concept is theoretically significant, it is hard to apply, and if it is easily applicable, too often it is theoretically trivial (Hull 1997: 358).

Hull's analysis is undoubtedly correct. If we are looking for a *single* species concept to function adequately relative to all three criteria, we will be disappointed. None seems to do so. A search for a satisfactory species concept on this strategy seems doomed.

THE SOLUTION

What I will be arguing here is that there are good reasons no single species concept has fared well against these criteria, and we should not expect one to fare well. We need to look elsewhere for a solution to the species problem. As important as these criteria are, the critical question is not about the adequacy of individual species concepts relative *to all three criteria*. The solution to the species problem is to be found at a higher level – at the level of the conceptual framework. The solution begins with the idea of a *division of conceptual labor*. Different species concepts function in different ways and should be judged on different grounds and against different criteria. Some concepts are theoretical and tell us what species are. Other concepts are operational and tell us how to observe species. What matters, then, is how a concept works within a particular framework, and how well it plays its appropriate role in the division of conceptual labor. As it turns out, there are reasons to maintain hope for a single theoretical concept, but still accept and even embrace a multitude of operational concepts. In this case, and perhaps others in science, a conceptual problem is solved not by looking at the adequacy of particular scientific concepts, but by looking at a particular *framework* of scientific concepts.

Part of the solution to the species problem will come from an account of theoretical concepts. I will not pretend to give a precise or full solution to the species problem here; that task is ultimately to be accomplished in the scientific debates. Nonetheless, one promising theoretical species is based on the idea of a *population lineage*. This is an idea that has been around for a long time, lurking even in the "essentialism" of the pre-Darwinian naturalists. It is also an idea that has become entrenched in both the attitudes of contemporary biologists and evolutionary theory. If we were to adopt the population lineage as the basis for a theoretical concept, we could then treat other species concepts as they are used – as operational concepts, or, in the terminology borrowed here from Rudolf Carnap, "correspondence rules."

13

A second component of the solution to the species problem comes from consideration of several different metaphysical stances – whether species taxa should be conceived more basically and generally as *sets* or *individuals*. Given what evolutionary theory tells us about species, are they sets of organisms with criteria for set membership, or are they concrete individuals with organisms as parts? As we shall see, this is not a question that we can answer directly and conclusively, nonetheless there are reasons to favor what has become known as the *species-as-individuals* thesis. This fundamental way of conceiving species coheres better with evolutionary theory, and provides more resources for developing and extending our thinking about species.

The third component to the solution comes from consideration of the functioning of individual species concepts. In order to more fully understand how species concepts work, we need to look also at a level lower than the overarching framework – at how individual species concepts get meaning, refer and change. We will start with a standard philosophical assumption that concept terms are "Fregean" – they have both meaning and reference. We will also look at the *definitional structure* of concepts and see how that helps us think more clearly about vagueness, referential indeterminacy and reference potential. Of particular relevance will be the roles and value of vagueness and referential indeterminacy.

Before we can turn to the solution, however, we need to better understand the species problem. One way to do this is by looking at the history of the species problem – the use of species concepts from Aristotle and Linnaeus through Darwin to the twentieth-century systematists. In chapter 2 we start this project, looking first at the ways Aristotle used the Greek term *eidos*, later translated as the Latin *species*. We then turn to the transformation of Aristotle by Porphyry, Boethius and others, and how this led to the medieval debates about the logic and nature of universals. Chapter 3 begins with the medical herbalists of the early Renaissance and their return to biology, then continues through the early naturalists, Ray, Linnaeus and Buffon, to the species problem Darwin confronted. Chapter 4 starts with Darwin's views – and doubts – about species, then continues through the development of species thinking we see in the architects of the Modern Synthesis – Dobzhansky, Simpson and Mayr, to the modern species debates.

Chapters 5 through 7 develop the three components of the solution to the species problem just outlined. In chapter 5 we start with the standard philosophical responses to the species problem – the nominalism that treats species as unreal, and the versions of pluralisms that cast doubt on the possibility of a single kind of species thing. We then turn to a more hopeful pluralism – the hierarchical pluralism of Richard Mayden and Kevin de Querioz that forms the foundation for the solution to the species problem. Chapter 6 addresses the second component of the solution – the competing metaphysical stances. We first look at different approaches to *species-as-sets*, contrast them with the *species-as-individuals* thesis, and evaluate them relative to evolutionary theory. Chapter 7 addresses the third component of the solution – the functioning of individual species concepts. Here we look at some standard philosophical assumptions about meaning, reference and definitional structure, to understand how species concepts function and change. Chapter 8 concludes the arguments here, beginning with a recap of the previous chapters, and an indication of where and how the analysis here might be useful beyond the narrow topic of focus here – the species problem. The species problem may seem to be just a problem for those who work in the biological sciences and the philosophy of science, but it has implications beyond – to conceptions of human nature, the functioning of scientific concepts in general, and an overarching metaphysical worldview that can help us think about change and the things that change.

Before we turn to the history of the species problem, clarity will be served by noting that the term *species* is ambiguous, referring to either the species category or to species taxa. Sometimes the specific use of *species* is clear from the context, but sometimes it is not. It is not clear, for instance, if Darwin's worries about the reality of *species* were about the species category or species taxa – whether there is a single kind of species thing or whether there are really groups of organisms that genuinely count as species. Because of this ambiguity, there are really two distinct but related questions. First, what is the species category? This is a question about species concepts. Second, what are species taxa? This is a question about the things we designate with the term *species*. This distinction is important. We might, for instance be realists about species taxa, but not about the species category – if we are pluralists and think there are multiple kinds of real species things. Or we might be essentialists about the species category without being essentialists about species taxa, if we think there is an essence to the category – a set of necessary and sufficient conditions for a group of organisms to count a species, but that species taxa don't in turn have essences – necessary and sufficient conditions for an individual organism to be a member of a species. The

15

solution developed here rejects the first stance, but endorses a version of the second – a *qualified* essentialism with regard to the species category, but a rejection of essentialism for species taxa. With this distinction in mind, we can turn to the history of the species problem, to understand how it came to loom so large in current thinking about species.

2

The transformation of Aristotle

THE ESSENTIALISM STORY

Modern species concepts are typically taken to have their origins in the essentialist frameworks developed by Plato and Aristotle that conceived species as having unchanging, eternal essences based on the possession of essential properties. This essentialist conception of species is usually assumed to have persisted long after Plato and Aristotle in the views of pre-Darwinian naturalists, Linnaeus in particular, until overthrown in the Darwinian revolution. This standard history, described by the historian of science Mary Winsor as "the essentialism story," is also the essentialism orthodoxy, given its widespread acceptance by philosophers and biologists. But at best it is misleading. It misrepresents Aristotle, Linnaeus and many, if not all, of the pre-Darwinian naturalists as being committed to an essentialism that groups organisms into species on the basis of essential properties or traits, and that implies species fixity, atemporality and discreteness. It also leads us to believe that Darwin's challenge was to a *property essentialism*. In this chapter, we will see where the Essentialism Story seems to go wrong with respect to the views of Aristotle and why. In the next chapter we shall examine the views of the pre-Darwinian naturalists, in particular Linnaeus, Buffon and Cuvier, and see how the Essentialism Story is misleading there as well. As we shall see, Darwin was not confronted with anything like the assumed essentialism consensus. But it is not just the history that suffers here. The Essentialism Story has led to a general misunderstanding of the species problem in general.

Most philosophers who discuss the history of species concepts start with the essentialism Aristotle supposedly inherited from Plato. And

17

they typically identify the essences of species with "essential properties." Daniel Dennett is one prominent example:

> Aristotle had taught, and this was one bit of philosophy that had permeated the thinking of just about everybody, from cardinals to chemists to costermongers ... [that] All things – not just living things – had two kinds of properties: essential properties, without which they wouldn't be the particular kind of thing they were, and accidental properties, which were free to vary within the kind ... With each kind went an essence. Essences were definitive, and as such they were timeless, unchanging, and all or nothing. (Dennett 1995: 36)

It is usually assumed that this form of species essentialism ruled until the time of Darwin. Dennett again:

> The taxonomy of living things that Darwin inherited was thus a direct descendant, via Aristotle, of Plato's essences ... We post-Darwinians are so used to thinking in historical terms about the development of life forms that it takes a special effort to remind ourselves that in Darwin's day species of organisms were deemed to be as timeless as the perfect triangles and circles of Euclidean geometry. (Dennett 1995: 36)

On another formulation, species essences are conceived in terms of unchanging "necessary and sufficient" properties that pick out all and only those individuals that are of a particular natural kind. Elliott Sober explains this general essentialist approach:

> *Essentialism* is a standard philosophical view about natural kinds. It holds that each natural kind can be defined in terms of properties that are possessed by all and only members of that kind. All gold has atomic number 79, and only gold has that atomic number. It is true, as well, that all gold objects have mass, but *having mass* is not a property unique to gold. A natural kind is to be characterized by a property that is both necessary and sufficient for membership. (Sober 2000: 148)

He describes species essentialism in these terms: "An essentialist view of a given species is committed to there being some property which all and only the members of that species possess" (Sober 1992: 250). Marc Ereshefsky agrees with Dennett in identifying the views of Plato, Aristotle and Linnaeus as essentialist and in terms of essential properties, but he also employs the terminology Sobers uses, "necessary" and "sufficient" properties (Ereshefsky 2001: 16–17). David Stamos similarly endorses the Essentialism Story, but looks as well to theological sources:

> Although species essentialism is a view which has (to put it mildly) lan-
> guished in recent decades, it has enjoyed a long and distinguished his-
> tory, being traceable back, broadly speaking, to the views of Plato and
> Aristotle on the one hand and the Book of Genesis on the other. The
> combination of these two traditions found its culmination in Carolus
> Linnaeus. (Stamos 2004: 22)

The main thesis of the Essentialism Story, that prior to Darwin species
were generally taken to be identified with some unchanging set of properties, is largely taken for granted among most philosophers and biologists.
This is in spite of doubts that have been raised by historians and historically minded philosophers about various details of the story (Balme
1987a, 1987b; Grene and Depew 2004; Lennox 1980, 1987a, 1987b, 2006;
Pellegrin 1987; Sloan 1985; Winsor 2003, 2006a, 2006b).

The details of this property essentialism are often left unstated, but
we can sketch out some of its most basic assumptions. First, the properties that can potentially be essential, or necessary and sufficient, are
usually assumed to be the kinds of properties we find listed in the works
of Aristotle and Linnaeus. In Aristotle we find properties such as being
cleft-footed or *blooded,* or *feathered.* In Linnaeus' taxonomic tables we
find properties such as the shape and number of stamen of flowers. These
properties are generally *intrinsic* in a standard philosophical sense. By
this it is usually meant that the presence of these properties does not
depend on anything external to the thing that has the property. *Extrinsic*
properties, on the other hand, depend on something external. (See
David Lewis 1983: 111–112, for one statement of this distinction.) The
properties of being *feathered* or *blooded* are intrinsic, because they do
not depend on anything external, whereas *being* a *sister* or *having an offspring* are extrinsic because they depend on a relation with an external
sibling or a parent. Often these essential properties are also assumed to
play an explanatory role (Sober 1992: 250). Being *feathered,* for instance,
might explain the locomotion and thermoregulation of birds.

Another standard assumption of property essentialism is that it rules
out species change because *the set* of essential, or necessary and sufficient, properties is unchanging. What counts as an essential property
cannot change, so if being *feathered* is an essential property for being
a bird, then it is always part of the essence – even if not all birds are
feathered in the same way and to the same degree. This implies that
species taxa are eternal. The set of properties associated with a particular species essence stays the same, whether or not any actual organisms
have these sets of properties. Furthermore, since species essences are

19

identified with particular sets of properties, different species must have different sets of essential properties. This suggests that species essences are distinct. On the standard *property* essentialism of the Essentialism Story then, species are immutable, eternal and discrete (Hull 1988: 82).

As the Essentialism Story also goes, Darwin vanquished this property essentialism by challenging the assumption that there is an unchanging set of essential properties associated with species taxa. He is taken to have accomplished this on the basis of two principles. The first is *population thinking*, which treats species as groups of organisms that vary within and across geographic locations. Variability within and among populations suggests that there is no single set of properties associated with all members of a species at any particular time (See Sober 1980: 351; 1992: 274). The second principle is a *gradualism* that treats species as lineages that change over time. Because of this change, there is no single set of properties associated with all members of a species over time (Ereshefsky 2001: 96). Combined, these Darwinian principles seem to imply that there is no single, definitive and unchanging set of essential properties to be identified with a particular species.

The Essentialism Story raises many questions: Are species ultimately *sets* or *classes* of things or are they something else? What does it mean to have a property and how should we understand the distinction between intrinsic and extrinsic properties? How are essential properties distinguished from the non-essential? If it is by virtue of the explanatory role of essential properties, then what constitutes such a role? How can there be degrees of a property that functions as an essence? I will return to some of these questions in later chapters, in particular, I shall ask whether species are sets or classes of things. But what is most important about the Essentialism Story is that it seems to be wrong. The figures most often identified as species essentialists, Aristotle and Linnaeus in particular, were not essentialists in the assumed *property* way. To see why, we will first examine the views of Aristotle. The views of Linnaeus must wait until the next chapter. In both cases, we will see how the Essentialism Story goes wrong, and misses important complications in the species problem.

ARISTOTLE AND THE LOGIC OF DIVISION

According to the Essentialism Story, Aristotle was engaged in the project of classifying organisms into genera and species on the basis of their respective essential properties. Usually this is taken to occur within the framework of the *logic of division* he developed in his non-biological works, the *Posterior Analytics* and *Metaphysics* in particular. The basic idea is that to determine the essence of a thing, we need to determine its generic nature and the attributes that differentiate it – the *differentiae* – from other kinds of things that share its generic nature. A species defin- ition would then give its essence: a list of properties that are necessary to be a member of a genus, and a set of properties that differentiate it from other species in the genus. Ereshefsky explains this standard construal of Aristotle: "The real essence of a species is a combination of its genus and the *differentia* that distinguishes it from other species of that genus. For example, the species man belongs to the genus animal, and its *differ- entia* is rationality. Hence, the real essence of a man is being a rational animal" (Ereshefsky 2001: 20). Aristotle's system is usually understood as hierarchical and *dichotomous*, the differentia being the presence or absence of a trait.

> According to the method of dichotomous division, a proper classification consists of a hierarchy of classes, each defined by the genus it belongs to, and its *differentia* within that genus. Furthermore, a genus should be dichotomously dividing according to which entities have a particular *dif- ferentia* and which do not. The genus of animate objects is divided into animals and vegetables according to the *differentia* of self-movement. Moving down one level in the hierarchy, the genus animal is divided into man and the lower animals according to the *differentia* of rationality. (Ereshefsky 2001: 20)

On this standard understanding, Aristotle was engaged in the project of classifying all things in nature into a hierarchy of genera and species based on the presence or absence of particular properties such as self- movement and rationality.

While this may be an accurate account of Aristotle's method of divi- sion, in his biological works he rejected that method for living creatures. In Book I of his *Parts of Animals*, he explicitly denied that the demarca- tion of kinds is based on dichotomy. "One should try to take animals by kinds, following the lead of the many in demarcating a bird kind from a fish kind. Each of these has been defined by many differences, not according to dichotomy" (Aristotle e643b9–12). And: "Some writers propose to reach the ultimate form of animal life by dividing the genus into two differences. But this method is often difficult, and often imprac- ticable" (Aristotle 642b5–7). Aristotle rejected dichotomous division for a variety of reasons. First is that many of the criteria we might employ in dichotomous division are superfluous. The differentiae *footed*, *two-footed* and *cleft-footed*, might all be employed, but only the last was regarded by Aristotle as significant (Aristotle 642b8–10). Second, "privative terms," based on the presence and absence of a trait, admit of no further subdivision. *Footless*, for instance does not admit of a further subdivision into the different kinds of *footlessness*.

> [P]rivative terms inevitably form one branch of dichotomous division, as we see in the proposed dichotomies. But privative terms admit of no subdivision. For there can be no specific forms of negation, of Featherless for instance or of Footless, as there are of Feathered or Footed. Yet a generic differentia must be subdivisible; for otherwise what is there that makes it generic rather than specific? There are to be found generic, that is specifically subdivisible differentiae; Feathered for instance and Footed. For feathers are divisible into Barbed and Unbarbed, and feet into Manycleft, and Twocleft. (Aristotle 642b22–30)

Third, and most importantly, many of the dichotomous attributes divide into unnatural groups (emphasis added):

> It is not permissible to break up a natural group. Birds, for instance, by putting its members under different bifurcations, as is done in the published dichotomies, where some birds are ranked with animals of the water, and other placed in a different class … If such natural groups are not to be broken up, *the method of dichotomy cannot be employed, for it necessarily involves such breaking up and dislocation.* (Aristotle 642b10–020)

Aristotle concluded his criticism here in the *Parts of Animals* with the outright rejection of dichotomy: "It is impossible then to reach any of the ultimate animal forms by dichotomous division" (Aristotle 644a12–13).

After criticizing dichotomous division, Aristotle gave us only rough guidance on how to proceed in determining which differentiae are helpful and satisfactory, suggesting that the larger groups are determined on the basis of general similarity and that the differentiae that demarcate subdivisions – species – are based on differences in the degree of attributes such as size.

> It is generally similarity in the shape of particular organs, or of the whole body, that has determined the formation of larger groups. It is in virtue of such a similarity that Birds, Fishes, Cephalopoda and Testacea have been made to form each a separate class. For within the limits of each such class, the parts do not differ in that they have no nearer resemblance

22

than that of analogy – such as exists between the bone of a man and the spine of a fish – but differ merely in respect of such corporeal conditions as largeness smallness, softness hardness, smoothness roughness, and other similar oppositions, or in one word, in respect of degree. (Aristotle 644b8–15)

Aristotle's failure to give a precise account of how to do this grouping may be due to his worries that it can be done unequivocally. And this may in turn be due to the lack of clear lines of demarcation as he explained in a famous passage (emphasis added).

Nature proceeds little by little from things lifeless to animal life in such a way that *it is impossible to determine the exact line of demarcation, nor on which side thereof an intermediate form should lie.* Thus, next after lifeless things comes the plant, and of the plants one will differ from another as to its amount of apparent vitality; and, in a word, the whole genus of plants, whilst it is devoid of life as compared with an animal, is endowed with life as compared with other corporeal entities. Indeed, as we just remarked, there is observed in plants a continuous scale of ascent towards the animal. (Aristotle 588b4–13)

But furthermore, many creatures also have equivocal attributes – resemblances to different classes such as animals and vegetables (emphasis added):

In the sea, *there are certain objects concerning which one would be at a loss to determine whether they be animal or vegetable.* For instance, certain of these objects are fairly rooted, and in several cases perish if detached; thus the pinna is rooted to a particular spot, and the razor-shell cannot survive withdrawal from its burro. Indeed, broadly speaking, the entire genus of testaceans have a resemblance to vegetables. (Aristotle 588b 13–17)

Significantly, here in Book I of the *Parts of Animals*, Aristotle did not give us guidance on how to resolve this conflict, nor is it clear that he tried to do so elsewhere. It might be, as David Balme concludes, that division is simply incapable of generating a hierarchical biological classification and Aristotle knew it:

[T]hey [differentiae] cannot form a hierarchical system because they cross divide. This was evident to Aristotle, for he points it out several times. At HA 1. 490b19–27, discussing the major genera, he shows how a neat scheme distinguishing *quadruped-viviparous-hairy* from *footless-oviparous-scaly* is upset by the viviparous footless viper and selachians, so that all that can be got out of these groupings is the non-convertible proposition that hairy animals are viviparous. (Balme 1987b: 84)

Instead of using his method of logical division to classify creatures, it seems that Aristotle just assumed conventional groupings in his *Parts of Animals*, as James Lennox argues:

An interesting feature of this extended criticism of dichotomous, arbitrary division and the development of a systematic, multi-differentiae alternative is that it takes for granted the 'naturalness' of certain kinds that 'the many' have identified. (Lennox 2006: 11)

In his *History of Animals,* Aristotle expressed a similar attitude (Lennox 2006: 15). The bottom line is that, while Aristotle appealed to animal examples in his non-biological works to illustrate the method of logical division, he did not use that method in his biological work to systematically classify organisms into species kinds. Nor did he seem to think that the method could be so used.

There are additional problems with the orthodox understanding of Aristotle as a species essentialist in the property sense. First, Aristotle did not use the Greek terms *eidos* and *genos* (translated as the Latin *species* and *genus*) in the systematic and biological way we use them today. These terms were not applied exclusively to living things, and they were not applied at set, fixed taxonomic levels. Instead they were applied at various levels. In his *Posterior Analytics,* for instance, Aristotle referred to geometry and arithmetic (in translation) each as a *genos* (Aristotle 75a38–40). Here he also gave a general account that apparently identifies a whole of *any kind* as a *genos*, and *any* subdivision as an *eidos* (species), providing examples in terms of numbers and geometric figures:

When you are dealing with some whole, you should divide the genus into what is atomic in species – the primitives – (e.g. number into triplet and pair); then in this way attempt to get definitions of these (e.g. of straight line and circle and right angle); and after that, grasping what the genus is (e.g. whether it is a quantity or a quality). (Aristotle 96b–20)

This process is applicable at different levels:

We should look at what are similar and undifferentiated, and seek, first, what they all have that is the same; next, we should do this again for other things which are of the same genus as the first set and of the same species as one another but of a different species from those. And when we have grasped what all these have that is the same, and similarly for the other, then we must again inquire if what we have grasped have anything that is the same – until we come to a single account. (Aristotle 97b7–13)

The idea is that – at any level of analysis – that which is similar and undifferentiated relative to some attribute or set of attributes will be a *genos*. And those things differentiated from others within a *genos*, but similar among themselves will be an *eidos*. Then we can regard this *eidos* as a genus and look for a further subdivision into *eide*. The process of division then operates at multiple levels.

In his biological works, Aristotle similarly used *eidos* (species) and *genos* (genus) in different ways and at different levels, generally relative to the investigation at hand. In a single passage from the *History of Animals*, for instance, Aristotle treated as *gene* the blooded animals, those quadrupeds who give live birth, and those who lay eggs, as well as fish, birds and cetaceans (Aristotle 505b 25–35). But elsewhere he treated quadrupeds and birds as *eide* (Aristotle 99b3–6). It is hard to see *genos* and *eidos* as each limited to a fixed, single level of grouping, and at a level that corresponds to modern taxonomic usage. (See also Lennox 1980: 324, fn. 11.) The only constant seems to be a *relational* one. In *genos-eidos* comparisons, the *genos* indicates a more inclusive level of grouping than *eidos*.

ARISTOTLE'S FUNCTIONAL ESSENTIALISM

Perhaps what is most important is that Aristotle seems to have been committed to an entirely different project. Instead of observing the attributes of animals in order to group them into animal kinds, he was more interested in discovering the general principles of the distribution of attributes or *parts* among organisms, and in order to give explanations.

> The course of exposition must be first to state the essential attributes common to whole groups of animals, and then to attempt to give their explanation. Many groups, as already noticed, present common attributes, that is to say, in some cases, absolutely identical – feet, feathers, scales, and the like; while in other groups the affections and organs are analogous. For instance, some groups have lungs, others have no lung, but an organ analogous to a lung in its place; some have blood, others have no blood, but a fluid analogous to blood, and with the same office. To treat of the common attributes separately in connexion with each individual groups would involve, as already suggested, useless iteration. For many groups have common attributes. (Aristotle 645b 1–14)

25

In this passage, when Aristotle used the term *group* he does *not* seem to have been referring to what we would identify as species taxa, but rather to groups of animals with particular traits such as feathers. What is important is not that particular kinds of organisms have a particular attribute, but *how* that attribute gets correlated with other kinds of attributes in *whatever kinds of organisms* they are found. Lennox explains:

> Aristotle appears to use the sentence form "As many as are X, all have Y" for a quite specific reason. It is not *just* to note a universal correlation; it is to do so while leaving the extension of the correlation open – a brilliant methodological innovation. New animals with the correlation can be discovered, but the generalization will not change, since it is about the universal correlation among differentiae ... not about the kinds that exemplify it. (Lennox 2006: 16)

This project is also apparent in his *Posterior Analytics*, where Aristotle used the coextension of traits to explain why vines shed their leaves.

> For let shedding leaves be *A*, broad-leaved be *B*, vine *C*. Well, if *A* belongs to *B* (for everything broad-leaved sheds its leaves) and *B* belongs to *C* (for every vine is broad-leaved), then *A* belongs to *C* and every vine sheds its leaves. *B*, the middle term is explanatory. But one can also demonstrate that the vine is broad-leaved through the fact that it sheds its leaves. For let *D* be broad-leaved, *E* shedding leaves, *F* vine. Well, *E* belongs to *F* (for every vine sheds its leaves) and *D* to *E* (for everything that sheds its leaves is broad-leaved); therefore vine is broad leaved. Shedding its leaves is explanatory. (Aristotle 98b6–16)

There are two critical points here: First, it appears to be an *explanatory* rather than a *classificatory* project that was motivating Aristotle (Lennox 2006: 8). The second critical point is that the *eidos/genos* (species/genus) framework was applied in Aristotle's biological works primarily to the *parts* of animals, not the animals themselves, as Pierre Pellegrin argues:

> [I]f one reads the biological corpus simultaneously putting aside taxonomic presuppositions and paying attention to the properties of the conceptual schema *genos-eidos*, one cannot fail to notice that this conceptual schema is applied preponderantly and fully not to animal classes but to the *parts* of animals. (Pellegrin 1987: 335)

He concludes: "Aristotelian biology is fundamentally a study of parts" (Pellegrin 1987: 335; see also Balme 1987b).

It is within this explanatory project based on the correlation of parts that Aristotle's essences are to be found. When he discussed *necessary conditions* and *essences* he typically did so in terms of functioning and development (emphasis added):

> The fittest mode, then, of treatment is to say, a man has such and such parts, because the essence of man is such and such, and *because they are necessary conditions of his existence,* or, if we cannot quite say this then the next thing to it, namely, that it is either quite impossible for a man to exist without them, or, at any rate, that it is good that they should be there. And this follows: because man is such and such the process of his development is necessarily such as it is; and therefore this part is formed first, that next; and after a like fashion we should explain the generation of all other works of nature. (Aristotle 640a 33–640b 5)

In this passage the term *necessary conditions,* refers to what we might describe as the *functional and developmental conditions* required for a certain kind of life, rather than the *logical conditions* of a taxonomic definition. What is *necessary* for being a thing is that the functional requirements for existence are met, and that development proceeds in a particular sequence – not that there are some set of intrinsic properties that satisfies a taxonomic definition. This is also an explanation. We are told that "a man has such and such parts," because these are needed for existence, and this enables us to understand why "the process of his development is necessarily such as it is." Lennox explains:

> By far the central question on Aristotle's mind throughout the discussion is *to what end* do these variations exist. Why is the hawk's beak hooked and why are its talons sharp and curve? Why are a duck's toes united by webbing, and why is its beak wide and flat? For what purpose are the legs and toes of cranes long and thin, and is there some relationship between these facts and cranes' relative dearth of tail feathers and wings? (Lennox 1980: 341)

This functional approach that focuses on differentiae does not – and cannot – produce a classification of organisms into species as implied by the Essentialism Story. It focuses on the *relation* between the parts of an organism, an environment, and way of life, not on the grouping of organisms on the basis of intrinsic, non-relational properties.

> [A]t least in regard to living things, the *essence/accident* distinction is a distinction between those features which are required by the kind of life an animal lives and those which aren't. If a crane is to survive and

flourish, it *must* have, not simply *long* legs, but legs of a certain length, defined relative to its body, neck length, environment, feeding habits, and so on. (Lennox 1987b: 356)

Aristotle did not see essences as the basis for species groupings in the modern sense. He saw them as the basis for understanding the functioning, development and flourishing of organisms in environments. Essential properties are necessary for a particular lifestyle in an environment. Accidental properties are not. This is clearly not the property essentialism orthodoxy usually attributed to Aristotle, for it does not involve using his logic of division to arrive at *definitions* of animal species kinds expressed in terms of their intrinsic properties.

To fully understand Aristotle's views, we need to distinguish three different senses in his use of the terms *eidos* and *genos*. One sense is logical and conceptual, what Phillip Sloan describes as a "logical universal":

> Understood as universals...Aristotle's concept of species and genera carry no particular biological significance. Any entity is subject to the predication of its *genos* and *eidos*, and there is no particular restriction of the usages to living beings. (Sloan 1985: 103)

On another sense, *eidos* is an immanent or "enmattered form":

> As enmattered form, *eidos* is not *per se* a universal or logical class, but constitutes the primary being of an empirically discernible entity. As form, *eidos* is both individual and non-material. In its pure individuality – as the immanent principle of shape, structure, intelligibility and order of a particular individual thing – it is not expressible in language, nor definable in itself. (Sloan 1985: 104)

The third sense of *eidos*, associated with the term *psuche*, is developmental. Here *eidos* is the "immanent principle of organization and vitality," as Sloan explains, contrasting it with the second, "enmattered form" sense:

> The clear identification of *eidos* with *psuche* in the *De Anima* introduces a further complication. Not only is the *psuche* related to body as form to matter by this identification; it also renders the *eidos* the dynamic principle of life, an immanent principle of organization and vitality which assumes the role of the formal, final and efficient cause of all organic activity. (Sloan 1985: 104)

On the first, logical universal sense, which we find in Aristotle's logical works, the relation of *eidos* to *genos* is hierarchical and worked out in terms of the method of division. On the second, enmattered form sense, which we find in Aristotle's biological works, the relation of *eidos* to *genos* is not hierarchical and is expressed in terms of the material form – the *eidos* – perpetuated in generation – the *genos*. On the third, developmental sense, the *eidos* is the principle of organization that governs growth and development.

These three senses of *eidos* were not new with Aristotle. He claimed that the first logical sense comes from the Pythagoreans through Plato (Peters 1967: 47). The second meaning, as enmattered form, has a longer history, appearing in the work of Homer and meaning "what one sees," or "appearance" or "shape of the body" (Peters 1967: 46). Corresponding to this sense of *eidos*, is an understanding of *genos* in terms of origins (as in "genesis"), and understood as a lineage, race or family. According to Sloan, this is this primary meaning to be found in Aristotle's biological works: "In this tradition the term *genos* tended to be the main "biological" application, designating a genetic "race" or family, without the more technical usage of it as classificatory *genos*" (Sloan 1985: 104).

The third sense of *eidos* as principle of organization, is perhaps the most distinctly Aristotelian, and in fact may have originated with Aristotle (Peters 1967: 50). On this third sense, the genus serves as the *matter* for the differentiation of *eidos*. It provides the system of shared characteristics that form the basis of the specific differentia (Lennox 1980: 336). Each specific difference is an actualization of some potential possessed by the *genos*. And the differences among the *eide* within the *genos* are to be understood and evaluated not by their form or structure, but by their suitability to a particular life in a particular context. The differences here are typically just a matter of degree – more or less rather than discrete, and the functional value is paramount:

> Viewed in abstraction from their specific environments, the variations between the beaks, wings, feathers, or legs of different birds appear to be nothing but variations in degree of the common differentiae of the genus of birds. But when viewed as adaptations to a peculiar mode of life, it becomes clear that each variation is the proper and peculiar one for a given species' way of life; no other differentiation would serve the needs of the species. (Lennox 1980: 342)

Because the variation among the species of birds is typically just a matter of more or less, beak and leg length for instance, and because the variations are based on the generic system of shared characteristics, the species within a genus will typically not have discrete morphological boundaries. There will be no clear demarcation between the species on the basis of particular intrinsic properties. The essential traits are essential because of the functional role they play in a mode of life, and this functional role is typically satisfied by differences of degree – beak length, leg length and so on. Accidental traits or properties don't play such a functional role. Lennox calls this the "teleological method" of distinguishing essential and accidental traits. The bottom line is that it is the functional unity of creatures as formed in development – the possession of functionally essential (necessary) traits relative to a particular lifestyle that makes them what they are:

> The "form" of living things is their proper activity, and the generic material is differentiated into a species member for the sake of this. What ensures the integrity and reality of Aristotelian species – in spite of the fact that their organs differ structurally only in degree along a variety of continua – is that the members of each species are individuals adapted to a specific manner of life. (Lennox 1980: 343)

On the basis of this, Lennox argues that "with its stress on the way in which the requirements of adaptation determine the precise differences between species within a genus, it is a spiritual ancestor of the evolutionary approach" (Lennox 1980: 345). Significantly, it is not obvious that Aristotle's views here preclude change. It may even be part of the functional essence for the parts to vary as the conditions of life require.

It would be wrong to see this ambiguity in *eidos* and *genos* as mere carelessness on Aristotle's part. His epistemology seems to require it. For Aristotle, knowledge of particular *eide* as enmattered forms is by sensation alone. Knowledge of the highest sort, however, depends on systematic logical deduction, and that requires the logical sense of *eidos*. As Sloan puts it: "To be knowable beyond the individual entity of sensation, it must be understood as universal" (Sloan 1985: 104). If we ignore the ambiguity, we miss a fundamental tension in Aristotle's epistemology – the tension between *being* as particular, and *knowledge* as universal. Recognition of this tension also allows us to see the tension in Aristotle's views about language – the predicates we use and how they relate to things in the world. Moreover, the distinction between enmattered form and principle of organization governing development is crucial in Aristotle's theory of explanation that distinguishes a formal cause from the *telos* or end. This multifaceted understanding of *eidos* is no careless mistake but is deeply embedded in his philosophy.

UNDERSTANDING THE ESSENTIALISM STORY

Whether or not we agree with Lennox in seeing Aristotle's functional essentialism as the "spiritual ancestor of the evolutionary approach," we surely have Aristotle wrong if we understand him just in terms of the orthodox Essentialism Story – as advocating a property essentialism that uses the method of division to divide and group organisms into species and genera on the basis of essential properties. First, he seems to have just assumed the commonly accepted classifications of organisms into plants, fish, birds, viviparous and oviparous quadrupeds and other groupings, rather than adopting classification as a goal. Second, his use of *eidos* and *genos* at multiple levels and outside the biological realm suggests he was not thinking about *species* in anything like the modern biological sense. Third, the description of various kinds of organisms was intended not for the classification of organisms, but for understanding the correlation of the attributes or parts of animals, which are the products of development and serve the functional requirements of a lifestyle. So when Aristotle discussed "necessary conditions" and "essences" in his biological works, it was relative to development and the functioning in an environment, not to biological taxonomy in the modern sense. His essentialism is functional and developmental in both focus and substance.

If all this is correct, how did we get Aristotle so wrong? As we shall see in the next chapter, part of the explanation is going to be found in the twentieth century, in the influence of Ernst Mayr – although Mayr himself did not think Aristotle to be an essentialist in the orthodox property sense. But any explanation of our misreading of Aristotle must presuppose the plausibility of the Essentialism Story. There are several reasons why this story is at least plausible. First, as we have already seen, Aristotle used *eidos* in more than one way. And an exclusive focus on the logical meaning would make Aristotle look *as if* he were a property essentialist. Since philosophers from late antiquity to the late medieval period, typically began and ended their study of Aristotle in the works on logic, it is no surprise that they read Aristotle in the manner of the Essentialism Story. But those who came later and read both the logical and biological works have also puzzled over the apparent inconsistencies. It might seem as if Aristotle were engaged in different projects in these different works. In his *Posterior Analytics*, for instance, he seems to be much more in line with the essentialism orthodoxy in his reliance on the method of division than in the biological works where he rejected that method and treated species as enmattered forms and principles of organization.

This apparent inconsistency has led some commentators, most famously Werner Jaeger, to attempt a resolution by assuming that Aristotle's views changed over time. Jaeger argued that Aristotle was a Platonist while he was at Plato's Academy, but became more empirical after leaving the Academy (Barnes 1995: 16–17). The logical works of the *Organon* – the *Categories* and *Posterior Analytics* in particular – would then represent his earlier a priori approach, while the biological works – the *Parts of Animals* and the *History of Animals* – would represent his later, more empirical approach. The earlier works might then plausibly endorse the use of division to classify, based on a lingering Platonism, while the later works might reject this use, based on the difficulties made apparent by empirical study.

That Aristotle used *eidos* in multiple ways suggests that there may be no need to postulate this developmental interpretation. But even if it were necessary, there are difficulties with Jaeger's "two Aristotles" explanation. First, the works we have that are attributed to Aristotle are undated. And there is little evidence to date them on other grounds, by references within the works, letters, etc. The few fragments that are plausibly dated to his early career are inconclusive about his views (Barnes 1995: 18–19). Even more problematic is the fact that Aristotle's works seem to have been rearranged by editors from the time that he left the Lyceum to the compilation of his works several centuries later by Andronicus. Andronicus organized Aristotle's writings according to topic rather than order of composition (Barnes 1995: 11–12). So parts of *each work* may well come from different periods.

I won't pretend that there are no exegetical problems here. There are too many questions about Aristotle's corpus in general. Nonetheless, recent scholars such as Balme, Pellegrin, Lennox and Sloan have shown that there is a coherent way to understand Aristotle's apparently inconsistent claims about *eide* (species). And they have made clear the problems with the Essentialism Story. But the careful analysis and insights of these scholars have been little noticed by nearly all of those who accept the Essentialism Story. David Stamos, however, is aware of these views, dubbing them the "new orthodoxy." He is also critical of this challenge to the standard Essentialism Story, claiming that "Aristotle thought that biological species constitute a distinct level of being and are themselves objectively real" (Stamos 2004: 108). He then asserts that "from the examples which he gives, such as 'sparrow, crane' … 'horse, man, and dog' … it becomes clear that what he had in mind is basically what we today think of as species" (Stamos 2004: 111). These two claims are highly problematic. First, Stamos seems to ignore the different ways *eidos* gets used in Aristotle's work. As we have seen, even within the *eidos-genos* hierarchy of the logical works, the term *eidos* gets used at all sorts of levels and to all kinds of things. There surely does *not* seem to be a unique level of being associated with *eidos*. Furthermore, it is implausible that Aristotle was generally using the term *eidos* to refer to those groups of organisms that we identify as species taxa. *Sparrow*, for instance, does *not* refer to a species taxon in our usage. In modern classificatory terms, *sparrow* is the family *Passeridae*, which includes multiple genera, and many species. The species level is instead constituted by *Passer ammodendri*, *Passer castanopterus*, *Passer diffusus*, *Passer flaveolus*, and so on. It is highly implausible that Aristotle could have had "in mind" the things we now identify as *species* taxa.

At the end of his discussion Stamos reveals in a striking passage, what seems to really be lurking behind his objection to the "new orthodoxy":

> Finally, when it comes to the traditional interpretation of Aristotelian species as essentialistic abstract classes, what must never be overlooked or underrated, it seems to me, is the sheer force and pervasiveness of that tradition. While the Medieval schoolmen added to Aristotle's ontology an individual form for each human … so as to allow for the immortality of individual souls, no-one in fame and influence did as much as the Swedish botanist Carolus Linnaeus (1707–1778), the self-proclaimed "Prince of Botanists" … to bring to fruition what may be called the Aristotelian paradigm. (Stamos 2003: 111)

The claim here is apparently that we need to interpret Aristotle as an orthodox essentialist because it is so much a part of the essentialist "paradigm," in particular as it is manifested in the views of Linnaeus. As we shall see, one problem with this argument is that the Aristotelian essentialist "paradigm," as Stamos understands it, is largely a fabrication of the twentieth century in the work of Ernst Mayr and others. Moreover, we shall see where it goes wrong with Linnaeus and other pre-Darwinian naturalists in the next chapters as well. But Stamos is not entirely wrong about this tradition. Some of those who followed Aristotle interpreted him as holding a view something vaguely like the essentialism orthodoxy I have been criticizing. In order to fully understand the orthodox essentialism story we need to look at how its seeds were planted in the period after Aristotle's death and through the Medieval debates about universals.

ARISTOTLE TRANSFORMED

After Aristotle's death, philosophical attention turned to the Epicureans, Stoics and Academic Skeptics. It is not clear precisely what happened to Aristotle's work. Evidence suggests that we have in our hands only about a third of what he had written. On one speculative account, Aristotle's colleague, Theophrastus, inherited it, handed it off to a nephew, Neleus, who then took it to Scepsis and hid it in a cave. These manuscripts were then discovered two centuries later, taken first to Athens and then to Rome, where Andronicus prepared an edition, organizing the works according to topic (Barnes 1995: 11). This edition is likely the source for modern compilations of Aristotle, as well as for those who were engaged in analysis and commentary from late antiquity on. Early commentators were interested in getting his views right, but also in developing and extending his views. Later commentators came from the Neo-Platonist tradition and were primarily interested in reconciling Aristotle's views with those of Plato. Christian theology came to influence the later commentaries as well, demanding that Aristotle's views be reconciled with Christian views about original sin, immortality of the soul and the doctrine of the Trinity. Most significant here, however, is the extensive debate that developed about the nature of "universals."

From the beginning of Aristotelian commentary, there was a focus on the *Categories*. This work was typically the first studied in the curriculum, and was the starting point of commentary, on the assumption that it provided the framework for his philosophy – in spite of the fact that there was no consensus about how to interpret it. The title itself comes from the Greek term for *predicate*. On a superficial reading it seems to be just about that – the use of predicates in language. Traditionally, the *Categories* has been divided up into three sections, the "Pre-Predicamenta," the "Predicamenta" and the "Post-Predicamenta" (after the Latin title of the entire work *Predicamenta*). In the Pre-Predicamenta, Aristotle first distinguished primary and secondary substances. Primary substances are the individual things to which the predicates apply, but are not themselves predicates. They are the subjects of predication, the things we talk about. "A substance – that which is called a substance most strictly, primarily, and most of all – is that which is neither said of a subject nor in a subject, e.g. the individual man or the individual horse" (Aristotle 2a13–15). Secondary substances are not individual things, but are that which is predicated of primary substances. Species and genera are therefore secondary substances. "For example, the individual man belongs in a species, man, and animal is a genus to the species; so these – both man and animal – are called secondary substances" (Aristotle 2a 15–17).

Aristotle then gave a division based on the ideas of "being said of" and "being in." Here, primary substances – individual things – are said not to be in a subject or to be said of a subject. Secondary substances, on the other hand, are not in the subject, but can be said of the subject. In other words, the predicates of a thing do not denote parts of it. Regarding the secondary substances *man* and *animal*:

> For man is said of the individual man as subject but is not in a subject: man is not in the individual man. Similarly, animal also is said of the individual man as subject, but animal is not in the individual man. (Aristotle 3a 8–15)

Like secondary substances, the differentia are not in the subject but are said of the subject: "For footed and two-footed are said of man as subject but are not in a subject; neither two-footed nor footed is in man" (Aristotle 3a 23–25).

It is through these predicates, which are substances only in a secondary sense, that we come to know the primary substances – the individual things.

> It is reasonable that, after the primary substances, their species and genera should be the only other things called secondary substances. For only they, of things predicated, reveal the primary substance. For if one is to say of the individual man what he is, it will be in place to give the species or the genus (though more informative to give man than animal); but to give any of the other things would be out of place – for example to say white or runs or anything like that. So it is reasonable that these should be the only other things called substances. (Aristotle 2b 29–36)

Analogous to the relation between primary and secondary substances is the relation between secondary substances (translated as the Latin *species* and *genera*) and other sorts of predicates – including the differentia.

> But as the primary substances stand to everything else, so the species and genera of the primary substances stand to all the rest: all the rest are predicated of these. For if you will call the individual man grammatical, then you will call both a man and an animal grammatical; and similarly in other cases. (Aristotle 3a 1–5)

The main section of the *Categories*, the Predicamenta, then provides a categorization of these other predicates that are not secondary substances. In the Predicamenta there is a ten-fold division of "things that are said." Aristotle began with the category *substance* itself. Then turns to quality, quantity, relation, place, time, position, state, action and affection. The Post-Predicamenta concludes with a discussion of related topics such as modes of opposition, priority, simultanaeity, change and "having."

There is currently no agreed upon interpretation of the *Categories*. Some see it as work about language and the patterns of predication. Others see it as about things and the nature of reality. Yet others see it as about the concepts that ground linguistic practices. This disagreement is not surprising. Aristotle shifts many times from locutions about "what is said" to "what is" without clarification or apparent recognition of the ambiguity. And, as we have already seen, the terms *eidos* and *genos* (which appear throughout this work) are systematically ambiguous in Aristotle's work. This disagreement about the interpretation of the *Categories* also has a long history. How it plays out in the commentaries on Aristotle's work has relevance to the understanding of the Essentialism Story.

The first group of commentators was "authentic" in the sense that they approached the works of Aristotle with the goal of getting him right. The last of these commentators of real significance, Alexander of Aphrodisias, was also widely regarded as the most important and the best of this group, and became known simply as "the Commentator." In the first century AD, he wrote commentaries on much of Aristotle's works, including the *Categories*, but not on the biological works (Frede 2009: 3). Nonetheless, he had some familiarity with the biological works, referring to them in other commentaries (Madigan 1994: 81). Like most of the commentators, he treated Aristotle's work as a unified, consistent, systematic whole. Two doctrines in Alexander's commentaries are most important for our purposes. First, is his recognition of the competing demands of being and knowledge, as Arthur Madigan explains: "Particulars are sensible, not intelligible; if all that exists is sensible, then knowledge is reduced to sensation which is absurd ... Certainly he and Aristotle would agree that knowledge is not to be reduced to sensation" (Madigan 1994: 84). Second, Alexander comments on Aristotle's different senses of *eidos* – not just as logical universal, but as enmattered form, and developmental or "productive" principle:

> Aristotle is right to hold that there must be an eternal εἶδος, eternal as matter is, but this does not have to be the εἶδος that comes to be in matter.

On the contrary, the ειδος that preexists is productive ειδος (the ειδος of the parent or agent or efficient cause), and this is like ... the ειδος which is produced. (Madigan 1994: 87)

Alexander's recognition of these different senses of *eidos*, and their epistemic significance gets lost in subsequent commentaries.

The Neo-Platonist commentators that followed had strikingly different goals. They were not so much interested in getting Aristotle's views correct, but in reconciling them with those of Plato. The story here might well begin in the third century with Plotinus, founder of the Neo-Platonist tradition, who read Aristotle as an especially valuable commentator on Plato, for his direct knowledge of Plato's views. But he did not assume to the degree of those who followed, that Aristotle was in agreement with Plato on the main philosophical issues. It is Porphyry, a generation later, who edited Plotinus' works and wrote *Life of Plotinus*, the work most responsible for the interpretive tradition that followed (Sorabji 1990: 17; Ebbesen 1990c: 141). First, he established the central importance of the *Categories* in the study of Aristotle. After Porphyry, and for many centuries after, the standard curriculum began with a study of the *Categories*, as an introduction to Aristotle (Blumenthal 1996: 22; Falcon 2005: 8). Then he took the work of Aristotle as a whole to be an introduction to Plato's philosophy. In this way, the works of Aristotle became the starting point for the discussions and commentaries on Plato. The assumption here was that, despite the apparent differences, the philosophies of Plato and Aristotle were really in agreement. The title of one of Porphyry's works makes this explicit: "On the School of Plato and Aristotle being One" (Sorabji 1990: 2).

Porphyry also wrote a comprehensive commentary on the *Categories*, and a simpler commentary for less advanced readers, that were to become the starting point for many of the commentators that followed (Falcon 2005: 8). His *Isogage*, or "Introduction," to Aristotle also came to serve as a general reference for many who followed. Porphyry's significance here is first, that after him there was long tradition of commentary that saw Aristotle and Plato as in harmony; second, he made Aristotle compatible with Plato by assimilating the views of Aristotle into those of Plato. In particular, he interpreted the *Categories* as being primarily about language and predication, and then assimilated it to a Platonic metaphysics based on a division of two realms, the intelligible and the sensible, with the priority of the intelligible. The subject matter of the *Categories*, according to Porphyry, was limited to the sensible realm, so what was found there was simply not able to contradict Plato's theory of the Forms. (Ebbesen 1990c: 145–146). Third, he set up what would be for the next millennium, the formulation of the "problem of universals" in his *Isogage*:

> At present ... I shall refuse to say concerning genera and species whether they subsist or whether they are placed in the naked understandings alone or whether subsisting they are corporeal or incorporeal, and whether they are separated from sensibles or placed in sensibles and in accord with them. Questions of this sort are most exalted business and require great diligence of inquiry. (Jones 1969: 186)

This particular formulation of the problem was adopted by the Neo-Platonist and Christian commentator Boethius.

Boethius, born near the end of the fifth century, took as his goal the translation of the entire available corpus of Aristotle from Greek into Latin. He failed to do this, but he did translate some of Aristotle's logical works and produced commentaries on the *Categories*, *On Interpretation*, and Porphyry's *Isogage* (Ebbesen 1990a: 374). These works, which relied heavily on Porphyry's commentaries (Shiel 1990: 350–351), would come to be the standard source for Aristotle throughout the Middle Ages, remaining important even after the rediscovery of Aristotle's texts from Arabic sources in the twelfth century (Marenbon 2005: 15). What is relevant here is that Boethius took Porphyry as his starting point for understanding Aristotle, accepting Aristotle's logic and Plato's metaphysics, and seeing them as part of the same philosophical project. His response to Porphyry's problem of universals makes this clear. By this time, commentators had come to understand Porphyry, in the passage just quoted, as claiming that universals could be interpreted in three ways: as concepts, as intrinsic to bodily things, and existing separately from bodies. Boethius seemed to accept this formulation of the problem of universals, arguing first, that universals subsist in sensible things, but are also understood apart from bodies; second, that they are not mere constructions of the mind but grasp reality as it is (Marenbon 2005: 3–4). And in his *Consolations of Philosophy,* Boethius seemed to endorse the Platonic view that there is a timeless reality independent of enmattered things (Hyman and Walsh 1983: 115). For Boethius, there was an isomorphism between the universal terms – the words, the logical sense of *species* and *genus* – as concepts, and the things in the world. And like other commentators, he understood the terms *species* and *genera* as universals and applying to non-biological categories as well as the biological. He treated *justice*, for instance, as a species of the genus *the good* (Hyman and Walsh 1983: 130).

The Neo-Platonist tradition played out also in more clearly theological ways. Augustine had interpreted Plato's metaphysics with a Christian gloss, where universals – species – were ideas in the mind of God. In his Trinity, he tells us:

> In Latin we can call the Ideas "forms" or "species," in order to appear to translate word for word. But if we call them "reasons," we depart to be sure from a proper translation – for reasons are called "logoi" in Greek, not Ideas – but nevertheless, whoever wants to use this word will not be in conflict with the fact. For Ideas are certain principal, stable, and immutable forms or reasons of things. They are not themselves formed, hence they are eternal and always stand in the same relations, and they are contained in the divine understanding. (Klima 2003: 197)

Lurking here is the idea that as reasons, *species* play an explanatory role in understanding the things in nature. They are the timeless principles that governed the creation, and served as the exemplars for all the individual enmattered forms that we observe, and that we use to derive via abstraction, the logical universals. This idea that the universals were ideas in the mind of God, employed in the creation gives a rationale to the isomorphism assumed by the NeoPlatonists of words, concepts and things. The ideas of God determine the things in the world, to which we apply our terms, and serve as the basis for the concepts in the understanding.

SPECIES AS UNIVERSALS

Historians generally agree that after Boethius there was a long gap before the return of significant philosophical thought in Europe (Marrone 2003: 15). It was only after the rediscovery of Aristotle in the twelfth century from Arabic sources that philosophy began to flourish in the monasteries and then universities of Europe. Nonetheless there were philosophical discussions in this period. One of the most important philosophical issues at this time was the relation between faith, revelation and reason. Lurking here was the question: how can Christians convince the Mohammedans and infidels of the truth of Christianity? Infidels and Mohammedans don't accept the faith and revelation that convinces those who already believe. Reason could play an important part here – if it were properly understood. It was generally assumed that in order to understand reason, an adequate account of the objects of reason, "universals," was required. (Lurking here also was the problem of the Trinity: how could three persons be really one?) The standard starting point for the Christian theologians of the Middle Ages was the "signification" of names: if the terms *Socrates* and *Plato* signified particular individuals experienced in sense perception, what do words like *man* and *animal* signify? The standard answer was *universals* – the species and genera of Aristotle's *Categories* (Jones 1969: 186). But what are universals? The answer given to this question was assumed to be within the framework laid out by Porphyry in his analysis of the three alternatives quoted above. Porphyry had asked first, whether species and genera subsist or are in the understanding alone; second, if they subsist whether they are corporeal or incorporeal; third, whether they are separate from or in sensibles. Historians have traditionally identified three main answers to Prophyry's questions: *realism, nominalism* and *conceptualism*. (This is clearly an oversimplification. For a criticism of this traditional history see Thompson 1995: 412, 418–419.)

The default Christian/Neo-Platonist position, based on the idea that universals are ideas in the mind of God, was realism. Universals exist independently of the particulars in which they are instantiated, because they exist timelessly in the understanding of the Divine mind and governed the creation of particular things. This was the view of many theologians, most notably of John Scotus Erigena in the ninth century, William of Champeaux and St. Anselm of the eleventh and twelfth centuries and Henry of Ghent in the thirteenth century (Jones 1969:187; Klima 2003: 199; Thompson 1995: 409–414). Realism had great theological value relative to the problems of the trinity and original sin. If universals were real, then *God* could refer to something real, be instantiated in the three persons of the Trinity – and yet still be "one." And if *man* referred to something real and independently of its instantiation in particular men, then something that happened to the universal *man,* such as the original sin, could still be true of each particular man (Jones 1969: 188). Some realists took this view to the extreme. Odo of Cambrai, for instance, claimed that the human soul is "essentially one," so that the creation of a child is not the creation of a new substance, but the creation of a new property in an existing substance – the human soul (Thompson 1995: 415). Other realists such as Anselm saw in universals a reflection of a divine hierarchy of increasing universality to God itself – the highest pure absolute (Thompson 1995: 414).

The early nominalists, on the other hand, adopted the opposite view that universals were mere words, and did not signify anything. Roscelin of Champiègne is typically taken to be the leader of this school of thought, arguing that universals were mere vocal utterances (Thompson 1995: 409, 419). Nominalism was clearly a problem for the standard theological commitments, if taken to imply that God the Father, the Son and the Holy Spirit were three different persons. On these grounds, at the Council of Soissons in 1093, Roscelin was ordered to repudiate his teachings (Jones 1969: 189). This doctrine also seemed to have the absurd implication that individual humans had nothing more in common than the spoken words that designated them.

The third position, usually described as "conceptualism," is most often identified with Peter Abelard (or Abailard) who argued in the latter half of the eleventh century for an Aristotelian "moderate" alternative to realism and nominalism. It is unsurprising that he should try to work out a moderate position, as he was a student of both the realist William of Champeaux and the nominalist Roscelin. The standard way of describing his views is in terms of concepts. Universals don't signify things of the world; they signify concepts in the mind. Abelard does not quite put it that way, however, where he lays out his position in his *Glosses on Porphyry*. Following Roscelin's example, Abelard starts with the idea that universals are words. After giving reasons to reject the realist claim that universals are things, he argued:

> Now, however, that reasons have been given why things can not be called universals, taken either singly or collectively, because they are not predicated of many, *it remains to ascribe universality of this sort to words alone.* Just as, therefore, certain nouns are called appellative by grammarians and certain nouns proper, so certain words are called by dialecticians *universals*, certain words *particulars*, that is, individuals. (Hyman and Walsh 1983: 177)

Universal terms are distinguished from particular terms by predication:

> A *universal* word, however, is one which is apt by its invention to be predicated singly of many, as this noun man which is conjoinable with the particular names of men according to the nature of the subject things on which it is imposed. A *particular* word is one which is predicable of only one, as Socrates when it is taken as the name of only one. (Hyman and Walsh 1983: 177)

A universal term then, is one that can be predicated of many individual things. *Man* for instance can be predicated of Socrates, Plato, Aristotle,

Boethius and so on, whereas each of these particular terms (proper names) can only be predicated of particular, single individual things. So far, Abelard could agree with the nominalists, but he was not satisfied to see it as nothing more than a matter of the usage of certain words. There is a psychological component in the process of abstraction, and a causal connection to the world in the "conjoining of predication."

According to Abelard, we arrive at "conceptions of universals" or "common nouns" by a process of abstraction and conjunction:

> In relation to abstraction it must be known that matter and form always subsist mixed together, but the reason of the mind has this power, that it may now consider matter by itself; it may now turn its attention to form alone; it may now conceive both intermingled. The two first processes, of course, are by abstraction; they abstract something from things conjoined that they may consider its very nature. But the third process is by conjunction. For example, the substance of this man is at once body and animal and man and invested in infinite forms; when I turn my attention to this in the material essence of the substance, after having circumscribed all forms, I have a concept by the process of abstraction. (Hyman and Walsh 1983: 183)

Abstraction consists in the selective attention to some subset of features of an individual thing. In individual humans, the form can be attended to without attention to the matter. And the faculty of reason can be considered apart from the animal nature.

> For, when I consider this man only in the nature of substance or of body, and not also of animal or of man or of grammarian, obviously I understand nothing except what is in nature, but I do not consider all that it has. And when I say that I consider only this one among the qualities the nature has, the *only* refers to the attention alone, not to the mode of subsisting, otherwise, the understanding would be empty. (Hyman and Walsh 1983: 184)

Each of these features attended to separately can then be conjoined in two ways, first, by likeness; second, into a single conception:

> Nevertheless, perhaps such a conception too could be good which considers things which are conjoined, and conversely. For the conjunction of things as well as the division can be taken two ways. For we say that certain things are conjoined to each other by likeness, as these two men in that they are men or grammarians, and that certain things are conjoined by a kind of apposition and aggregation, as form and matter or wine and water. The conception in question conceives things which are so joined to

each other as divided in one manner, in another conjoined. (Hyman and Walsh 1983: 184)

Because there is a common cause to this conjunction, this is no mere nominalism. The qualities and how they are abstracted and conjoined is caused by the nature of the things themselves. Because Socrates *and* Plato had certain features in reality, the sharing of these features – the similarity – is the cause for the conjunction in the first sense. The cause for the conjunction in the second sense is the possession of the individual person, Socrates *or* Plato, of two of the features considered in abstraction – matter and form, reason and animal nature.

Universal conceptions for Abelard then have a dual nature. They are incorporeal in that they exist only in the understanding, but they are corporeal in that they exist in universal terms – spoken and written words, as well as the common cause of the features abstracted and conjoined, and as well as the signification of all the individual things – all the individual men. This comes out in Abelard's revisiting of the three alternatives laid out by Porphyry: whether genera and species subsist; whether they are corporeal or incorporeal; and whether they are in sensibles or not. In answer to the first question, Abelard concluded that species and genera in fact subsist because "they signify by nomination things truly existent, to wit, the same things as singular nouns" (Hyman and Walsh 1983: 186). In response to the second question, Abelard claimed that they are in a certain sense corporeal "with respect to the nature of things and incorporeal with respect to the manner of signification" (Hyman and Walsh 1983: 187). Similarly, with regard to the third question: species and genera "are sensible with respect to the nature of things, and that the same are insensible with respect to the mode of signifying" (Hyman and Walsh 1983: 187).

It is within this framework, that Abelard tried to reconcile the views of Plato and Aristotle. First, he contrasted the views of Plato and Aristotle on universals (or at least what Boethius describe as their views):

> Wherefore Boethius records that Aristotle held that genera and species subsist only in sensibles but are understood outside them, whereas Plato held not only that they were understood without sensibles but that they actually were separate. (Hyman and Walsh 1983: 171)

Then he argued that Plato and Aristotle can be reconciled.

> For what Aristotle says to the effect that universals always subsist in sensibles, he said only in regard to actuality, because obviously the nature

which is animal which is designated by the universal name and which according to this is called universal by a certain transference, is never found in actuality except in a sensible thing, but Plato thinks that it so subsists in itself naturally that it would retain its being when not subjected to sense, and according to this the natural being is called by the universal name. That, consequently which Aristotle denies with respect to actuality, Plato, the investigator of physics, assigns to natural aptitude, and thus there is no disagreement between them. (Hyman and Walsh 1983: 183)

While it is not entirely clear how to interpret this passage, the idea seems to be that the universal terms such as species and genera are associated with common concepts, that are then to be understood as the forms, as Abelard suggested: "For what else is it to conceive forms by nouns than to signify by nouns" (Hyman and Walsh 1983: 183). The common concepts – as forms – are not themselves sensible and so retain their "being" when "not subjected to sense."

Abelard's solution to the problem of universals is usually seen as Aristotelian. It is easy to see why. The Neo-Platonists typically reconciled Aristotle's views with Plato's, but either by making Aristotle's metaphysics more Platonic, or by restricting Aristotle's views in the *Categories* to language, and not about metaphysics. Then he could be seen as simply not commenting on Plato's theory of the transcendental forms. But Abelard turned Plato's transcendental forms into concepts of the mind, which were seemingly compatible with Aristotle's metaphysics. Abelard's conceptualism, however, was not the final word. In the century after Abelard, much of Aristotle's work was rediscovered through the translation of Arabic texts. In 1160, Abelard's student John of Salisbury began working out the "subtle" and difficult ideas in Aristotle's *Posterior Analytics*, although it would take another half-century before the text received any written commentary (Marrone 2003: 33). From this time on, analysis need not be limited to just the *Categories* and *On Interpretation*, based on the commentaries produced by Boethius and Porphyry. This rediscovery of Aristotle saw new attempts to interpret him within a Platonic framework, but it was nonetheless traumatic, with efforts to ban Aristotle's work, and all discussion of his work, occurring in 1210 and 1215. But by the 1240's Aristotle's work had been absorbed into the academic curriculum, albeit with condemnations of some theologically "dangerous" principles in 1277, 1284 and 1286 (Marrone 2003: 35).

This was the environment in which Thomas Aquinas found himself. In the middle of the thirteenth century, he fought these efforts to suppress

Aristotle's views by trying to work out an interpretation of this newer, more complete Aristotle that would be compatible with Christian theology. Aquinas was clearly committed to a strong form of realism, based on the idea that universals were ideas in the Divine understanding. But instead of knowledge of these ideas coming from Divine illumination, as it did for Augustine and Anselm, this knowledge came from an active human intellect capable of illumination on its own. As in Abelard's approach, this came about through a process of producing a conception by abstraction from a thing itself, as Aquinas explains: "When we speak about an abstract universal, we imply two things, namely the nature of the thing itself, and abstraction or universality" (Klima 2003: 201–203). Aquinas believed that God gave humanity a faculty, the intellect, that enables it to abstract from the "sense likenesses" of particular things the unchanging essential elements that make it the thing it is (Brown 1999: 253).

This realism is in contrast to the views of William of Ockham, developed a century later, and usually described as a nominalist. Ockham argued for a hierarchy of language, from the written, to the spoken, and then ultimately to a mental language that underlies the written and spoken. This is a relation of subordination with the written and spoken words ultimately dependent on the mental. In his *Summa Totius Logicae*, he started with what he took to be the Aristotelian view, as laid out by Boethius:

> According to Boethius in the first book of the *De Interpretatione*, language is threefold: written, spoken and conceptual. The last named exists only in the intellect. Correspondingly the term is threefold, viz. the written, the spoken and the conceptual term. A written term is part of a proposition written on some material, and is or can be seen with the bodily eye. A spoken term is part of a proposition uttered with the mouth and able to be heard with the bodily ear. A conceptual term is a mental content or impression which naturally possesses signification or consignification, and which is suited to be part of a mental proposition and to stand for that which it signifies. (Hyman and Walsh 1983: 653)

What is important is first that, for Ockham, universal terms, as spoken and written, are ultimately dependent on the universal terms of the mental language – the concepts. The term *man*, for instance, may be just a name, written or spoken, but it relies on a name – a concept – in the mental language for its meaning. Second, written and spoken language are conventional, but mental language is not. Mental terms or concepts "naturally signify" what they are concepts of – the things in nature they are predicated of. While for Ockham this mental concept is not a universal *thing*, which he thought to be an incoherent notion, it is universal in that it is predicated universally to all particular, individual men (Spade 2008: 14). Ockham therefore seems to be committed to a type of conceptualism, rather than the nominalism of Roscelin that identified the universal term as a mere vocalization. For this reason, some historians now see Ockham as a conceptualist rather than a nominalist (Hyman and Walsh 1983: 649). Other historians, on the other hand, suggest that Ockham is really a type of realist, a "nominalist realist," because the natural signification of universal terms is based on real similarities in nature, which are prior to any activity of the mind. *Man* signifies all those things that are human, because all these things really are alike, and not just because they are thought alike, or called the same thing in written, spoken or mental language (Brown 1999: 272).

After Aristotle there seem to have been at least five approaches to universals. First was the Platonic realism, which identifies universals with transcendent forms. Second was the Theological realism, which identified universals with ideas in the mind or understanding of God, that served as templates in the Creation. Third was the nominalism of Roscelin, which treated universals as mere spoken words. Fourth was the conceptualism of Abelard, which identified universals with concepts of the mind derived from the abstraction of the properties of particulars. Finally, there was the conceptualism or "nominal realism" of Ockham, which identified universals as particular terms of the mental language, on which spoken and written language depends, but which can be predicated of many things. On each of these approaches, the *eidos* and *genos* of Aristotle – and the corresponding *species* and *genus* of the Latin translations – are understood first, as *general* universal terms, not just *biological* universal terms; second, as primarily functioning within a context of language and logic; and third, as requiring consistency with theological metaphysical assumptions about the nature of God, creation and original sin. There can be little doubt that the Aristotle of the *Parts of Animals* and the *History of Animals* had been radically transformed.

CONCLUSION

At the beginning of this chapter, we addressed one of the central commitments of the orthodox Essentialism Story: Aristotle and many of those who followed were advocating a *property essentialism* that classified organisms into species using the method of division, which grouped organisms on the basis of the possession of essential, or nec- essary and sufficient, properties. We then saw how this Essentialism Story misrepresents the views of Aristotle. First, while Aristotle may have used his method of division in his logical works, in his biological he explicitly rejected its use to group animals into species kinds – *eide*. He just assumed the conventional kinds. Second, the terms *eidos* and *genos*, translated as *species* and *genus*, were used at multiple levels, and in non- biological contexts as well as biological. Aristotle did not use *eidos* at the single *biological* classificatory level we see in modern taxonomy. Third, he used the term *eidos* in at least three senses – as logical universals, as enmattered form, and as principles of development and organization. In his biological works, *eidos* typically referred to enmattered form or to principle of development and organization. But it is in the logical uni- versal sense that the method of division gets applied. Finally, the essen- tialism he embraced was not a classificatory *property* essentialism but a *functional-developmental* essentialism. Essences here are not the prop- erties that generate a classification, but those that are necessary for life – the development and functional correlation of parts. Aristotle's primary concern was not the grouping of organisms into species kinds, but the understanding of how hearts, lungs and other parts functioned in what- ever kinds of creatures had them, relative to particular modes of life.

The failure of the orthodox interpretation of Aristotle, the Essentialism Story, seems to demand an explanation. How did we get him so wrong? Part of the answer to this question is surely found in the interpretive tradition extending from the last of the "authentic commentators," Alexander of Aphrodisias, through the fourteenth century. One of the most distinctive features of this interpretive tradition is its reliance on a very narrow range of Aristotle's work. From at least Porphyry on until the rediscovery of Aristotle's work in the twelfth century, the main texts were the *Categories* and *On Interpretation*, with little knowledge of any of Aristotle's other works – especially the biological texts. Moreover, the understanding of the *Categories* and *On Interpretation* relied heavily on the analysis of two Neo-Platonists, first in Porphyry's commentaries and his *Isogage*, and second in the commentaries of Boethius, which then constituted the basis for the philosophical discussions of Aristotle until the twelfth century. Because both Porphyry and Boethius had limited access to Aristotle's work and both were focused on rendering Aristotle compatible with Plato, we should not expect an authentic account of Aristotle's views from either. And even with the rediscovery of the rest of

Aristotle's corpus, little attention was paid to his biological works. This is unsurprising, given the thousand-year focus on just a limited portion of his logical works, and the preoccupation with issues about meaning. It is also unsurprising given the theological focus of medieval philosophy. For the Christian theologians from Boethius to Abelard and Aquinas, it was crucial that Aristotle be rendered consistent with Christian theology. It was this focus that transformed the debate about the *eidos* and *genos* of the *Categories* into the *species* and *genus* of the Latin translations and commentaries, and then into the more general debate about universals. For many Christian commentators, *species* and *genera* – as universals – became ideas in the mind of God. Those who rejected this idea risked the wrath of those protecting the orthodoxy.

Earlier in this chapter, I addressed David Stamos' suggestion that part of the reason to treat Aristotle as an essentialist in this sense is the long tradition of interpreting him so. It is not difficult to see how the medieval commentary tradition might incline us to the Aristotle of the Essentialism Story. First, it is in Aristotle's logical works that the method of division is most at home. This is where *eidos,* as logical universal, is determined by division from the *genos* on the basis of differentia. Second, the absorption of questions about the *species* and *genus* of the *Categories* into the more general problem of universals reinforced the tendency to see it as a linguistic matter, about the nature and signification of common nouns or universal terms. This naturally led to an emphasis on definition, particularly in the works of Aquinas and Ockham. Finally, whereas Aristotle was devoted to the careful observation of nature, in order to understand the principles of development and the details of enmattered form, there was little value placed on observation in the time from Porphyry to the fourteenth century. For both the Neo-Platonists and Christian theologians true reality was not found in the enmattered forms, but in the transcendent forms and the ideas of God that served as exemplars in the Creation. Because of this emphasis and interpretive bias, the tradition missed much of what was interesting in Aristotle, in particular his worries about using division to classify animals, his complex understanding of the terms *eidos* and *genos*, and his focus on a functional-developmental essentialism. By getting him right on these matters, we will have the conceptual resources to better understand the debate about species that followed.

3

Linnaeus and the naturalists

THE ESSENTIALISM STORY REDUX

Those who understand Aristotle's project to be an essentialist system that classified organisms into unchanging species kinds on the basis of essential properties typically see a similar essentialism in many pre-Darwinian naturalists. In a passage quoted at the beginning of the previous chapter, Daniel Dennett claims that: "The taxonomy of living things that Darwin inherited was thus a direct descendant, via Aristotle, of Plato's essences" (Dennett 1995: 36). Here Dennett seems to be implying that, from Aristotle to Darwin, the property essentialism of the Essentialism Story was pervasive and predominant. Marc Ereshefsky describes this standard view:

> [T]he common philosophical stance towards species was essentialism. Linnaeus, John Ray, Maupertuis, Bonnet, Lamarck, and Lyell all adopted an essentialist (or typological) view toward systematics ... On this view, classification systems highlight natural kinds – groups whose members share kind-specific essences ...According to biologists who adopted an essentialist view of species, species have the same role in biological taxonomy as the chemical elements have on the periodic table. All and only the members of a species taxon should have a common essential property. (Ereshefsky 1992: 188–189)

Of the group mentioned by Ereshefsky, the Swedish botanist Carolus Linnaeus is perhaps the most prominent and is specifically seen as working in this essentialist tradition. Typically Linnaeus is also taken to have adopted Aristotle's method of division. Ereshefsky again:

> Linnaeus used the method of logical division as the foundation for his method of classification. An organic species is distinguished by its

49

differentia from the other species in its genus ... Yet the "real distinction," or essence, of a species cannot be given without the definition of its genus ... Thus we have the Aristotelian notion of a species' essence: its generic definition plus its species' differentia. Linnaeus also distinguishes those characteristics that are essential for membership in a species (whether they be part of that species' real essence or its necessary properties) from those that are accidental. (Ereshefsky 2001: 201–202)

David Stamos interprets Linnaeus as this sort of essentialist:

[F]or Linnaeus, raised on one of the last strongholds of Aristotelian scholasticism in Europe, the University of Uppsala in Sweden ... species are essentialist kinds, created from the beginning and remaining essentially unchanged ... defined binomially by a genus and differentia (which was a shorthand for a full description of essential characters) (Stamos 2007: 137)

This is at least plausible. After all, Linneaus developed and used the familiar binomial nomenclature that gives a genus-species name to taxa (*Homo sapiens* for instance), along with an apparently defining list of attributes associated with all members of a genus and a set of differentia that distinguishes one species from another. And he seemed to be treating some traits at least *as if* they were essential traits of species and higher taxa (Larson 1968: 299). There are nonetheless significant problems with this understanding of Linnaeus' views. As we shall see, he can be seen as a species essentialist in this sense only at the beginning of his career. But before we turn to the views of Linnaeus, it will be helpful to our understanding of his views to look briefly to the period preceding him, and following the rediscovery of Aristotle's works in the late twelfth century.

RENAISSANCE NATURALISTS

There was little study of nature in the period preceding and immediately after the rediscovery of Aristotle in the twelfth century. While there was a tradition of compiling illustrations of the various creatures of nature together in the "beastiaries" of the late medieval period, this tradition was not grounded on the careful observation of nature. Rather, beastiaries included mythical beasts along with actual, and descriptions were primarily undertaken for the purpose of illustrating particular religious morals, dogmas and values (Ogilvie 2006: 97). Bestiaries were ultimately fables based on religious dogma rather than studies in natural history. The "Cocodryllus" of the Nile, for instance, described as about thirty feet long, with horrible teeth and claws, was presented primarily for its moral heuristic:

> Hypocritical, dissolute and avaricious people have the same nature as this brute – also any people who are puffed up with the vice of pride, dirtied with the corruption of luxury, or haunted with the disease of avarice – even if they do make a show of falling in with the justifications of the Law, pretending in the sight of men to be upright and indeed very saintly. (Ogilvie 2006: 102)

There were at this time, however, groups of medical herbalists, whose studies of plants were associated with the apothecaries, and were instrumental in the production of medicines. This tradition seems to have been largely independent of, and unconcerned about, the philosophical debates about universals and the nature of species (Ogilvie 2006: 97–98). Perhaps because of this independence, the medical herbalists played a role in the development of the natural history tradition of the Renaissance, and ultimately to the modern systems of biological classification.

The natural history tradition began in the middle of the fifteenth century in Italy, and took the texts of Aristotle, Theophrastus, Dioscorides and Galen as starting points. The first workers in this tradition were reformers, looking to fix the errors of medicine, in particular errors in the formulation of medicines from plants by the apothecaries. The most important of this first group was likely Niccolo Leoniceno (1428–1524), who taught moral philosophy and medicine at the University of Ferrara (Ogilvie 2006: 30). Leoniceno, like many others in this tradition, began with the works of the ancients. He possessed a large library, with many Greek manuscripts, including some of Aristotle's biological works in both Greek and Latin translation (Ogilvie 2006: 31). Leoniceno assumed that some of the ancient texts were correct, but others were corrupted. One way to correct these errors was to look closely to the texts and the translations themselves. But another way to correct them was through the observation of nature. So for Leoniceno and his fellow medical herbalists, the starting point was the texts of Aristotle, Dioscorides and others, but that eventually led to the careful attention to nature.

Around 1530 there was a second group of herbalists that included Otto Brunfels, Hieronymus Bock and Leonhart Fuchs, as well as the Italian Pietro Andrea Mattioli. Like the earlier group, these herbalists were practical in orientation. They were not describing plants for the sake of mere description or classification, but for the sake of improving medicine (Ogilvie 2006: 36). Conrad Gessner (Konrad Gesner) was also important, by virtue of his massive *Historia Animalium* written over a seven-year period from 1551 to 1558 (Ogilvie 2006: 38). This group was more focused on the description of plants and animals than the first, but still retained a medical stance, and a starting point in the ancient texts. An ever increasing reliance on observation can be seen, though, in the "collectors" such as Carolus Clusius who followed, who gathered and cataloged plants in botanical gardens and herbaria (Ogilvie 2006: 39). This practice naturally turned the herbalists to more theoretical questions: how should an herbarium or garden be organized? And why? The larger the collection, the more important these questions became. The answers to these questions seemed to require something like a classificatory scheme and theory.

The next group of naturalists, at the end of the sixteenth century and the beginning of the seventeenth, started to do just that. In this period, the descriptions and illustrations became more detailed and precise. But perhaps more importantly, there was as yet no single naming scheme. A particular plant might be known by one naturalist under one name, and known by another under a different name. This problem became even more compelling given the explosion of newly named plants. The Lyon herbal of 1587–88, for instance, described more than 2000 plants (Ogilvie 2006: 48). Caspar Bauhin (or Gaspard Bauhin), who was critical of this particular herbal, devoted more than thirty years to reading through the various botanical publications of the previous hundred years, comparing descriptions and names with specimens in his own herbarium, trying to resolve the conflicting names and descriptions of plants. As a result of his efforts, plant descriptions became more precise and concise, and they became more informative in terms of distinguishing one kind of plant from another (Ogilvie 2006: 48).

Up until this time, classifications were typically in alphabetical order. Gessner mostly used this method in his *Historia animalium*. Beyond its practical mnemonic value, however, it seemed arbitrary – especially if there were multiple names for individual plants that would have them classified in different locations. Other methods though, seemed equally arbitrary, based on such subjective criteria as "plants with flowers that please," and "odorous plants," and ecological criteria such as "plants that are found in shadowy, wet, damp, and rich places" (Ogilvie

2006: 216–217). But by this time, there had also been some efforts to classify on the basis of similarity. Hieronymous Bock, Valerius Cordus and Leonhart Fuchs all *tended* to group similar plants together (Larson 1971: 9–11). This led to an increased effort to be more systematic and hierarchical. Bauhin, for instance, began sectioning plants into genera, each containing particular species. Some of these genus sections even seem to match modern classifications based on common descent. But crucially, he did not use an explicit species concept, and often seemed to just assume the commonsense folk taxa that divided the animal and plant worlds into kinds of things. Others of this period adopted similar approaches, separately classifying trees, shrubs and herbs, and dividing them into species (Ogilvie 2006: 216–219).

The most important naturalist of the time was likely Andrea Cesalpino (1519–1603) who brought together a number of different threads into a coherent, systematic approach. He explicitly advocated an Aristotelianism in his *De plantis libri XVI* of 1583 and in his *Questionum Peripateticarum* (Olgivie 2006: 209). He was Aristotelian first in his epistemology, arguing that knowledge consists in the predication and hierarchy of universals (Larson 1971: 24). He was also Aristotelian in his focus on the features that are central to the functioning of plants – the nutritive and reproductive features. These were to be the basis of division. James Larson explains:

> Division rests upon essential parts, that is, the major functional parts. In plants there are two functions – nutritive and reproduction – hence two systems of organs – vegetation and fructification. Within each of these systems Caesalpino calculates the relative importance of each organ by means of an a priori notion of finality. When, for example, we consider the reproductive function, it is clear that what constitutes reproduction is the production of seed, and accordingly seed plays an important role in division, while the other reproductive parts, performing subordinate roles, have a subordinate importance in division. (Larson 1971: 27)

Cesalpino's system was also hierarchical, and therefore consistent with Aristotle's views in his logical works about *eidos* and *genos* as logical universals – yet inconsistent with Aristotle's rejection of division in his biological works. But although it was hierarchical, it was still a hierarchy only of species and genera – not of the kingdoms, families, classes and orders that would later appear in the Linnaean system (Ogilvie 2006: 224). But for Cesalpino, as for the other naturalists of the period, a classificatory system also needed to satisfy some purely practical requirements. It must be easy to learn and apply. This required that it employ traits that were easy to recognize and that generated easily memorized classifications (Larson 1971: 38).

The main features of Cesalpino's system were used by the systematists that followed, in particular Robert Morison, Augustus Rivinus, Josef Pitton de Tournefort and John Ray (Larson 1971: 32). But there were also some important changes. First, while this later group continued to classify plants into species and genera, there was an increasing tendency to use these terms at fixed taxonomic levels. Instead of using the term *species* at multiple levels, depending on the particular level of inquiry, they began to reserve that term for the lower levels of classification. John Ray (1627–1705), for instance, adopted a hierarchical system, based on species and genera, but with a relatively fixed set of levels, from the most general *genera* (he also used the term "order" here) to the less general *species subalternae* (sometimes referring to these as "tribes") and the least general *species infimae* (Ray 1735: 21). But like Aristotle, Ray also used the term *species* to refer to non-biological things. In his *The Wisdom of God Manifested in the Works of Creation,* he speculated about whether all metals were of the same species or not, and the number of species of indivisible particles or atoms (Ray 1735: 60–61). Given this broader usage of the term *species*, it is easy to see why there was relatively little attention to the nature of biological species. Since minerals and atoms were also *species infimae*, we shouldn't expect a distinctively *biological* theory of species.

This group of systematists also followed Cesalpino in relying heavily on reproductive organs as "essential" traits. They did so on the same justification – that reproduction has Aristotelian functional significance, in that plants have souls with nutritive and reproductive capacities. On the basis of this, we might be tempted to think of them as the property essentialists of the Essentialism Story. There is something to be said for this interpretation. While they were taking features that were *functionally* essential, and were therefore following Aristotle in his functional-developmental essentialism, they were using these traits in way that Aristotle did not. They were using them as *taxonomically* essential, to generate species and genus groupings in the biological sense. That said, they were not generating their species and genus grouping *solely* on the basis of the essential reproductive traits. Like Cesalpino, they also employed operationally practical traits – traits that were easy to recognize, highly variable and that produced easily memorized classifications. All of these early naturalists wanted a system that had heuristic value – in assisting the naming, learning and memorization of all the many newly described plants and animals. There was then a tension between the theoretical and practical demands.

Sometimes the practical demands prevailed, and species and genus groupings were based on non-essential, "accidental" traits. For this reason, these naturalists typically admitted that their classifications were "artificial." But another reason that they regarded their classifications as artificial was based on the Aristotelian idea of the continuity in nature. The famous passage from Aristotle's *History of Animals,* quoted in chapter 2, asserts such a continuity: "Nature proceeds little by little from things lifeless to animal life in such a way that it is impossible to determine the exact line of demarcation, nor on which side thereof an intermediate form should lie" (Aristotle 588b4–13). A truly natural system would reflect all the many differences in nature, and the continuity in these differences. Most naturalists of the time, John Ray in particular, read this passage as denying the practical possibility of a truly complete and consistent natural system. James Larson explains:

> In relation to plant form, the natural method, when perfected, would admit no exceptions, and would be independent of the practical interest or value imputed by human will. Comprehending all plant parts, properties, faculties and qualities, such a method would consider roots, stems, and leaves, as well as flowers and fruits, and draw from the comparison of their resemblances and differences the affinities which resolved them in groups. Such a method is not, as John Ray says, the task of one man or one age. The idea of continuity, however, did have one immediate effect upon classification; it tended to make the whole notion of a hierarchical system appear a convenient but artificial division of natural forms, with no counterpart in nature. (Larson 1971: 42)

There is a tension here as well. On one understanding of Aristotle, a classification should be based on the functionally essential traits – reproductive in particular. But on another understanding, what was important was that a natural classification be based on all traits – in all their complexity and continuity – not just the essential. This view is also Aristotelian. Aristotle had denied that the method of division could be applied to generate animal kinds, precisely *because* of this complexity among living things.

During this time, there was also a theoretical turn to the thinking about species and an attendant concern about the species *concept.* Ray speculated in his *Wisdom of God* that many plants commonly regarded as species were really just "accidental varieties," but did not say there how to distinguish mere varieties from true species (Ray 1735: 24). He did, however, give an account in 1668 in his *Historia Plantarum*, based on reproduction (emphasis added).

> In order that an inventory of plants may be begun and a classification of them correctly established, we must try to discover criteria of some sort for distinguishing what are called "species." After a long and considerable investigation, no surer criterion for determining species has occurred to me than the distinguishing features that perpetuate themselves in propagation from seed. Thus, *no matter what variations occur in the individuals or the species, if they spring from the seed of one and the same plant, they are accidental variations and not sufficient to distinguish a species.* (Mayr 1982: 257)

This is a striking claim, and seems to conflict with what we might have expected – a criterion based on the possession of essential traits. For Ray, genealogy and reproduction were the true criteria for species membership, not the possession of essential properties.

What is most important in the views of the naturalists who followed Cesalpino is that first, the species category – *species infimae* – increasingly came to occupy a fixed taxonomic level, instead of operating at multiple levels, depending on the inquiry. Second, *functional* essential traits were given special theoretical status in classification, appearing to generate species groupings on the grounds asserted by the Essentialism Story – essential properties. But because of the use of practical traits, and because of the belief that *all* traits had to factor in a natural classification, the grouping principles were *both in practice and theory* heterogeneous. Finally, for Ray at least, species membership was determined by reproduction and genealogy not similarity. This is not the property essentialism of the Essentialism Story.

LINNAEUS

Perhaps the central figure in the Essentialism Story is Carolus Linnaeus (or Carl von Linné), born in 1707, two years after the death of Ray. Linnaeus adopted and developed the basic approach to classification that we saw first, with Cesalpino, then with Ray. He used sexual or "fructification" characters to group organisms, but he also employed practical, operational traits. He explicitly adopted a hierarchical system, and then expanded the classificatory scheme to include *kingdoms, classes* and *orders.* He restricted *species* to a fixed level of classification, although he also continued to apply the term outside the biological realm, to minerals in particular. And he systematically used the familiar binomial nomenclature, giving species the familiar genus-species name. But most significantly, we see with Linnaeus an increased interest in the theoretical issues related to biological species – what they are, whether they can change, and whether they can be created or destroyed. What he concluded seems to contradict the Essentialism Story.

According to the Essentialism Story, Linnaeus used Aristotle's method of division to classify organisms into species on the basis of essential, or necessary and sufficient, properties. This is usually taken to imply that, for Linnaeus, species taxa were unchanging, eternal and discrete. As the *set* of defining essential traits is timeless and unchanging, so must be the species taxa defined by them. The first problem with this story is that only at the beginning of his career did Linnaeus accept anything like the standard essentialist assumption that species were fixed, discrete and timeless. In 1735, while Linnaeus was studying medicine in the Netherlands, he published his first important work, the *Systema Naturae Sive Regna Tria Naturae Systematice Proposita per Classes, Ordines, Genera & Species.* In this first edition of only fourteen pages, he laid out the framework of his taxonomic system – three kingdoms of nature – mineral, plant and animal – each divided into classes, orders, genera and species. In the first of his "Observations on the Three Kingdoms of Nature," he asserted that no new species are produced: "Hinc nullae species novae hodienum producuntur" (Linnaeus 1964). And in the fourth observation, he seems to assert the fixity of species on the ground that "like always gives birth to like." Since offspring closely resemble parents, and since members of species tend to multiply over time, he inferred that there must have been some single original pair or individual of each species created by God, analogous to the creation of the original human pair – Adam and Eve.

But in 1742, seven years later, he encountered a specimen of the flower *Linaria* that also had distinct attributes associated with *Peloria*. This convinced him that new species could arise through hybridization. The apparent offspring species of these two kinds of flowers produced seeds that developed into fertile offspring that also resembled the parents in various ways. Linnaeus took this to be evidence of change, not just of degree but in kind (Larson 1968: 293; Eriksson 1983: 94–95). In 1744 he defended his conclusion in his *Dissertatio botanic de Peloria* and later described his conclusions in a letter to Albrecht von Haller:

> I beg of you not to suppose it [the *Peloria*] anything else than the offspring of *(Antirrhinum) Linaria*, which plant I know well. This new plant propagates itself by its own seed, and is therefore a new species, not existing from the beginning of the world; it is a new genus, never in being until now. (Larson 1968: 294)

While this did not turn out to be a true case of speciation, it did draw Linnaeus' attention to hybridization, which he believed to be confirmed in other cases. In a letter from 1751 he wrote:

> I find hybrid plants more common than hybrid animals and rather many in number. I believe I have got [been allowed] to open the door to one of nature's extensive chambers, although it is not opened without creaking. (Larson 1968: 295)

Linnaeus incorporated this discovery into later editions of his *Systema Naturae*. In the tenth edition of 1758, he wrote that God created an original individual or mating pair for each genus and that new species were produced by inter-generic crosses. And in the thirteenth edition of 1770, he speculated that the original breeding pairs or individuals might instead represent *orders*, rather than *genera*, and that even new genera, as well as species, might be formed through hybridization.

> We may suppose God at the beginning to have proceeded from simple to compound, from few to many! and therefore at the beginning of vegetation to have created just so many different plants, as there are natural *orders*. That He then so intermixed the plants of these orders by their marriages with each other; that as many plants were produced as there are now distinct *genera*. That Nature then intermixed these generic plants by reciprocal marriages ... and multiplied them into all possible existing species (Larson 1968: 297)

Significantly, he declined to speculate about whether there were limits to this creative process (Larson 1971: 107).

Linnaeus took this hybridization theory seriously. At least thirty-four years before the thirteenth edition, in his 1746 *Sponsalia plantarum*, he was working on a mechanism to explain how new species could appear through hybridization. In 1760, in his *De Sexu plantarum*, he worked out his ideas most clearly (Larson 1971: 107). Inspired by Cesalpino, who was in turn influenced by Aristotle, he theorized that there were two main elements: a medulla that was the essential bearer of life and associated with the female; and a cortex, that was the "conveyer of nourishment" and associated with the male (Larson 1971, Lindroth 1983: 44; Stafleu

1971: 137). The medulla was more essential and less variable than the cortex, as James Larson explains:

> In the cortex-medulla hypothesis plant fructification is part of the medulla descending from the mother and plant vegetation is part of the cortex descending from the father; plants issuing from the mixture of two different plants are similar to the mother in fructification and belong to her group although they resemble the father in outward appearance. The medullar substance, because "essential," is more or less the same in all plants, for it is by means of this substance that they and their offspring are plants ... The Cortical substance, however, because less essential is more variable ... The medulla when covered with the cortex consisting of a mixture of different principles constitutes a prototype for each natural order, which He then mixed to produced the progenitors of the genera; these genera when mixed one with another produced as many species as exist presently. (Larson 1968: 297)

Notice here that this is not just a theory of the formation of new species, but it is one that assumes common ancestry – the first breeding pair representing the originating order or genus!

The details of this medulla/cortex theory are less important for our purposes here than the following. First, shortly after asserting the fixity of species in the first edition of his *Systema Naturae*, Linnaeus seemingly abandoned this commitment. For at least thirty years he was instead committed to the view that new species and even new genera could be formed out of old. He believed in a dynamic process of species formation, and saw it as a natural process *governed by law* and with no assumed limits, as his son explained in a letter written shortly after Linnaeus' death:

> He certainly believed that the species of animals and plants, as well as the genera, were of time, but that the natural orders were the deeds of the Creator ... If Our Lord first created orders among animals and plants, and that gradually, He, in His omnipotence, and without destroying His laws, could certainly have allowed them to mix among themselves, and from these genera appeared; and leave them thereafter to the laws, He had given nature and implanted in every growing thing eventually to mix and produces species; how long this can continue it is not for us to seek to know. (Larson 1968: 298)

Second, if he was an essentialist, it was of a *genealogical* kind. Essences were passed on in reproduction via the transmission of medullar matter. But he was clearly not a *property* essentialist in the standard philosophical sense, because neither the medulla nor the cortex were associated with a particular set of intrinsic properties. Rather the intrinsic properties each organism has are a product of *the mix* between the more essential medulla and the more variable cortex. A single medulla could have different outcomes in terms of physical attributes depending on the nature of the particular cortex it was associated with in sexual reproduction. Most significantly, this combination may be a product of two different genera or orders. In other words, whatever essences there were, were not associated with any particular set of physical, intrinsic properties! Third, in Linnaeus's genealogical essentialism, an organism was a member of a species not because of a set of properties, but because of genealogy. Lurking here is the idea that species are lineages – groups of organisms connected not by the possession of similar traits, but by parentage and reproduction. Finally, lurking here also is a functional essentialism. *Essential* traits for Linnaeus, as for Cesalpino, were traits necessary for life – for the nutritional and reproductive functioning of organisms. This echoes Aristotle's use of the term. Essential traits, for Aristotle, were not taxonomic – used to determine inclusion into biological taxa – rather they were functional, related to the demands of life for a particular form with a particular mode of living.

If the essentialism of the Essentialism Story is committed to the view that species taxa are associated with an unchanging set of intrinsic properties and are therefore fixed, timeless and discrete, it is hard to read Linnaeus as being an essentialist in this sense. But by his use of binomial nomenclature to name kinds and list of descriptive traits to identify them, Linnaeus may look like an essentialist committed to Aristotle's logic of division. In his taxonomic tables, he *could have* been doing something like giving definitions for fixed, unchanging and discrete species. But his views about the origins of species paint a very different picture, a picture of species formation from an apparently unlimited process of mixing that constantly produces new forms out of old, and forms not necessarily well-defined and discrete. These are not fixed and timeless species. So is it possible to reconcile the implied property essentialism of his binomial nomenclature and taxonomic system with his views about the formation of new species through mixing?

The answer to this question is perhaps to be found in Linnaeus' conflicted understanding of his own classificatory project. He believed he was fulfilling a duty to understand God by uncovering the secrets of Creation in the observation and classification of all things in nature. For Linnaeus a fully natural classification correctly represents nature in all its complexity, as reflecting the ideas of God in the Creation. But Linnaeus also believed that he was far from achieving such a natural classification. First, there were many animals and plants that were simply unknown to naturalists. Only when these gaps could be filled would his classification even come close to representing God's plan (Larson 1971: 70). Sten Lindroth explains both the importance of a *natural* system to Linnaeus, and this inadequacy of his system:

> That system explored the extent of the Creator's plan link by link in the chain of the plant kingdom – the innumerable groups of plants in the spontaneous interrelatedness provided by nature. As clearly as anyone else, Linnaeus saw the urgency of his task. It is well known how time and again from his youth onward he insisted that a "natural" plant system was the real goal of botany, "the first and the last that should be sought in botany" ... But Linnaeus considered, rightly it might be thought, that in practice nearly insuperable difficulties were encountered in the attempt to carry out this program; the whole enterprise demanded a knowledge of all the species of plants on earth, which not even he himself possessed. (Lindroth 1983: 21)

Linnaeus did not think his system "natural" because it did not represent the full complexity of Created nature.

But there was a second, and more important reason he did not think his classification to be natural. In a natural system, each animal and plant had to be placed in a species, genus, order and class on the basis of *all* the features – all the similarities and differences – that revealed God's plan. The sexual, fructification traits he used were only provisional, and served as mere substitutes for the full range of traits a natural system required, as Lindroth explains: "It was always his view that his own sexual system was only a provisional tool, a *substitute for the natural method.* But further he would not go" (Lindroth 1983: 21). A truly natural system would not just be based on the "essential" traits, but on all traits, and in all their complexity.

> Linné conceived a natural method which represent all natural affinities fundamental in botany. All knowledge of natural objects depended upon the distinction of like from unlike. Such knowledge was, therefore, in proportion to the number of "real" or "natural" distinctions encompassed. The greater the number of distinctions a method could comprehend, the clearer became the ideas of natural objects. (Larson 1971: 62)

The point here is that the "essential" features Linnaeus used to generate his classification into species and higher levels were only provisionally valuable, as an *initial* attempt to represent the full complexity of God's plan. They were *not* understood by Linnaeus to be the ultimate basis for a natural classification. And they were *not* essential in a definitional sense, as providing a set of necessary and sufficient conditions for being a kind of thing. They were provisional criteria given in the hope that they could eventually lead to a natural classification.

There are then several tensions in Linnaeus' views about classification. First is the tension between practical value and theory. From the earliest Renaissance herbalists on, an important goal of classification was to facilitate first, the easy identification of plants (and animals); second, the production of a system that allowed for memorization. These practical demands required the use of easily observable and identifiable traits that varied in obvious ways. Plants, and animals to a lesser degree, were often classed together in species and genera on the basis of the traits that were observable, identifiable and variable, instead of the functionally essential traits that were necessary to nutrition and reproduction. The second and related tension is between the logical features of the classification, and the observational – or "reason and experience" as James Larson puts it (Larson 1971). A logical, systemic approach was easy to remember, and easy to use, but could not fully represent all of the continuities and complexities in nature. This, as we might recall from chapter 2, echoes Aristotle's reservations about the method of division in his biological works. In his *Parts of Animals* we found the outright rejection: "It is impossible then to reach any of the ultimate animal forms by dichotomous division" (Aristotle 644a12–13). In short, a logical system was, for Linnaeus, unnatural. Nonetheless, he favored the logical, hierarchical system, at least partly for practical reasons. This was the focus of criticism by his contemporary George-Louis Leclerc, the Comte de Buffon, who worried Linnaeus was promoting logic over reality, using a system that was logical, systematic and easy to apply, but that did not reflect the full complexity of nature (Lindroth 1983: 29).

BUFFON AND HISTORICAL SPECIES

This tension between logic and observation was also found in Buffon's thinking, but Buffon seemed to emphasize the complexity of nature over the simplifying demands of the logic, leading many to interpret him as a species nominalist. He seemed to hold something like a nominalist position in the first of his thirty-six volumes of the *Histoire Naturelle*, written from 1749 to 1788, where he affirmed a principle of continuity and then suggested that any divisions are arbitrary:

For in order to make a system or arrangement, everything must be included, and the whole must be divided into different classes, these classes into genera, and the genera into species – all this according to an order in which there must necessarily be something arbitrary. But Nature proceeds by unknown gradations, and consequently cannot wholly lend herself to these divisions – passing, as she does, from one species to another species, and often from one genus to another genus, by imperceptible shadings; so that there will be found a great number of intermediate species and of objects belonging half in one class and half in another. Objects of this sort, to which it is impossible to assign a place, necessarily render vain the attempt at a universal system. (Lovejoy 1968: 90)

A few pages later, he asserted that individuals alone exist:

In general, the more one increases the number of one's divisions, in the case of natural products, the nearer one comes to the truth; since in reality individuals alone exist in nature, while genera, orders, classes, exist only in our imagination. (Lovejoy 1968: 90)

One would think, given these passages, that Buffon would have rejected any species criterion. If there were only individuals, how could there be a criterion for deciding species membership? But that is not the case. In the second volume, published the same year as the first, he argued for a reproductive criterion:

We should regard two animals as belonging to the same species, if, by means of copulation, they can perpetuate themselves and preserve the likeness of species; and we should regard them as belonging to different species if they are incapable of producing progeny by the same means. Thus the fox will be known to be a different species from the dog, if it proves to be a fact that from the mating of a male and female of these two kinds of animals no offspring is born; and even if there should result a hybrid offspring, a sort of mule, this would suffice to prove that fox and dog are not of the same species – inasmuch as this mule would be sterile. (Lovejoy 1968: 93)

He took this reproductive criterion seriously enough to engage in a series of experiments. In 1755, he reported:

We do not know whether or not the zebra can breed with the horse or ass; whether the large-tailed Barbary sheep would be fertile if crossed with our own; whether the chamois is not a wild goat; whether the differences between apes are really specific or whether the apes are not like dogs, one species with many breeds ... Our ignorance concerning these questions is almost inevitable, as the experiments which would settle them require more time, care and money than can be spared from the fortune of an ordinary man. I have spent many years in experiments of this kind, and will give my results when I come to speak of mules. But I may as well say at once that I have thrown but little light on the subject and have been for the most part unsuccessful. (Lovejoy 1968: 95)

Even though these experiments were unsuccessful, the fact that he thought they could establish species distinctions is significant. It was not the presence of essential traits that established species membership, but the ability to interbreed.

For Buffon, this reproductive criterion allowed for the possibility of change. In the fifth volume of *Histoire Naturelle* (1755), he addressed the possibility of evolutionary change.

Though species were formed at the same time, yet the number of generations since the creation has been much greater in the short-lived than in the long-lived species; hence variations, alterations, and departures from the original type, may be expected to have become far more perceptible in the case of animals which are so much farther removed from their original stock. (Lovejoy 1968: 100)

This is not a true evolutionary change, such as we would see in Darwin. Change is constrained by an unchanging and permanent "moule intérieur." Nonetheless, Buffon thought change could be significant. In 1778, he wrote:

Though Nature appears always the same, she passes nevertheless through a constant movement of successive variations, of sensible alterations; she lends herself to new combinations, to mutations of matter and form, so that today she is quite different from what she was at the beginning or even at later periods. (Lovejoy 1968: 104)

This echoes an earlier passage in volume VI from 1756, where Buffon emphasized the role of environment:

If we consider each species in the different climates which it inhabits, we shall find perceptible varieties as regards size and form; they all derive an impress to a greater or less extent from the climate in which they live. These changes are made slowly and imperceptibly. Nature's great workman is Time. He marches ever with an even pace, and does nothing by leaps and bounds, but by degrees, gradations and succession he does all these things; and the changes which he works – at first imperceptible – become little by little perceptible, and show themselves eventually in results about which there can be no mistake. (Lovejoy 1968: 104)

What is striking here is that Buffon is emphasizing something that the Essentialism Story identifies with Darwin, the gradual change over time.

There are a number of important themes in Buffon's views. The first is a distinction, between logic – the abstract logical orders of nature imposed by the mind, and the real order *in* nature. Buffon thought Linnaeus to be far too concerned with the former. "This manner of thinking has made us imagine an infinity of false relationships between natural beings ... It is to impose on the reality of the Creator's works the abstractions of the mind" (Sloan 1976: 359). Linnaeus may have recognized that his system was not *natural*, because it did not reflect the full complexity of nature, but he nonetheless thought it to be a valuable, albeit provisional system. Buffon, in contrast, thought the abstraction in Linnaeus' system to be misleading and counter-productive. So while both Buffon and Linnaeus worried about the tension between logic and observation, they disagreed about its implications for classification.

The second and related theme in Buffon's work is that knowledge – true understanding – comes not from these mere logical, abstract orders, but from the real orders found in nature. We can see this in Buffon's general criticism of mathematical abstraction:

In all these cases one always makes suppositions contrary to nature, depriving the subject of the majority of its qualities, making of it an abstract being which no longer resembles the real being. One then transports this ideal result into the real subject. This is the most delicate and important point in the study of the sciences – to know properly how to distinguish that which is real in a subject from that which we have arbitrarily placed there in considering it. (Sloan 1979: 117)

Buffon criticized Linnaeus on the basis that he confused the abstract orders with the real orders. Sloan explains:

This concern with the separation of the *abstract* from *real* orders can then be clearly seen to underlie Buffon's attack on taxonomy, particularly Linnaean taxonomy. His argument is that Linnaeus committed the fundamental error of confusing the abstract and arbitrary order of concepts with the real and physical order of nature. (Sloan 1979: 117)

Because of this confusion of the abstract with the real, Buffon thought that the Linnaean system could not be the basis for true knowledge: "[The

Linnaean system] is not science, and is at most only a convention, an arbitrary language, a means of understanding; from it results no real knowledge" (Sloan 1979: 117).

The third theme in Buffon's work is that true relations in the organic world are found in the historical succession of individuals that reproduce and not in the arbitrary abstract relations of similarity. It is in this sense we can understand an early statement of his species concept in 1749: "This power of producing its like, the chain of successive existences of individuals, constitutes the real existence of the species [*existence réele*]" (Sloan 1979: 117). And in 1753 a similar claim in his article on *The Ass*: "It is neither the number, nor the collection of similar individuals which makes the species. It is the constant succession and uninterrupted renewal of these individuals which constitute it" (Sloan 1979: 117). This historical approach is not limited to species. Higher-level taxa, such as genera and families, were also conceived by Buffon historically, as groups of interconnected lineages. And subspecies were similarly lineages subordinate to species lineages (Sloan 1979: 118). In Buffon then, we find a comprehensive, historical understanding of biological taxa as lineages and groups of lineages.

The point here is that it is difficult, if not impossible, to see how Buffon could be interpreted unequivocally as a property essentialist in the manner usually understood of the pre-Darwinian naturalists. He seemed to think that the approaches based on similarity were mere arbitrary abstractions, did not reflect nature, and were therefore incapable of producing true understanding. True understanding was achieved through knowledge of real relations in nature, in particular the historical succession of interbreeding organisms. This is clearly not a property essentialism that groups organisms into species on the basis of an unchanging set of essential or necessary and sufficient *intrinsic* properties. Of course, this is not to say that Buffon regarded shared traits or similarities as irrelevant. Because offspring *tended* to look like parents, members of species *tended* to be similar in some ways. Similarity could then be relevant, but only insofar as it was a guide to the *real*, historical relations of succession. More pertinent was reproduction, which was necessary for *there to be* a series of individuals that could comprise a species. For Buffon, species were the historical series of individual organisms that were generated through reproduction – and recognized through ability to reproduce. This turn to the historical relations among individual organisms is highly significant. It is not new with Buffon however. Recall that this was one of Aristotle's ways of thinking about *eidos* – as form perpetuated in reproduction from a *genos* as "family."

This historical conception of species – as a succession of individuals – was adopted by other naturalists. In Diderot's *Encyclopédie* of 1755, for instance, there was a discussion of a "natural history" conception of species (Sloan 2003: 235). And Cuvier later endorsed a genealogical criterion (emphasis added):

> We imagine that a species is the total descendence of the first couple created by God, almost as all men are represented as the children of Adam and Eve. What means have we, at this time, to rediscover the path of this genealogy? *It is assuredly not in structural resemblance.* There remains in reality only reproduction and I maintain that this is the sole certain and ineffable character for the recognition of species. (Mayr 1982: 257)

Kant also adopted Buffon's historical conception, following his lead in distinguishing "logical division" from what he calls a "natural division," and giving priority to the latter:

> The logical division [of taxonomists] proceeds by classes according to similarities; the natural division considers them according to the stem [*Stämme*] and divides animals according to genealogy, and with reference to reproduction. One produces an arbitrary system for the memory, the other a natural system for the understanding [*Verstand*]. The first has only the intention of bringing creation under titles; the second intends to bring it under laws. (Sloan 1979: 128)

The laws Kant refers to in this passage include the "law of reproduction," that he credited to Buffon:

> In the animal kingdom the division of nature into genera [*Gattungen*] and species [*Arten*] is grounded on the general law of reproduction, and the unity of genus is nothing else than the unity of the generative force, which is considered as generally active for a determined manifold of animals. Thus, the Buffonian rule – that animals which can generate fertile young and which might show differences in form [*Gestalt*], belong to one and the same physical genus [*Physichen Gattung*] – can properly be applied only as a definition of a natural genus [*Naturgattung*] of animals generally, to differentiate it from all logical genera [*Schulgattungen*]. (Sloan 1979: 127)

Referring to this passage, Sloan argues that Kant followed Buffon in connecting reproduction to a historical conception of species: "the key to classification in the history of nature is not character resemblance, but the recognition of groups united by the historical unity of the stem" (Sloan 1979: 127). Sloan then quotes Kant on the difference between "logical" and "natural" division:

> The logical division [of taxonomists] proceeds by classes according to similarities; the natural division considers them according to the stem [*Stämme*], and divides animals according to genealogy, and with reference to reproduction. One produces an arbitrary system for memory, the other a natural system for the understanding [*Verstand*]. The first has only the intention of bringing creation under titles; the second intends to bring it under laws. (Sloan 1979: 128)

Like Buffon, Kant emphasized the distinction between "logic" and genealogy, and gave priority to the latter.

Some of those who followed Kant, such as Johann Georg A. Forster and Johann Blumenbach, remained committed to a similarity-based species concept, even while endorsing a reproductive criterion (Sloan 1979: 130–138). Others, such as Christoph Girtanner, endorsed the Buffon-Kant historical conception of species. In his 1796 "On the Kantian Principle for Natural History" he worked out Kant's arguments into a comprehensive plan for a new system of classification:

> The division of the animal kingdom which has been given here, and which has been taken from the most famous describers of nature [*Naturbeschreibern*], will likewise be useful for the historian of nature [*Geschichtschreiber*] until, through exact observations, trials, and experiments, the laws of reproduction are sufficiently known. Then, however, a new division of the animal kingdom into classes, orders, species [*Gattungen*], races, sports [*Spielarten*], and varieties must be undertaken, according to the relationship of reproduction. (Sloan 1979: 140)

Girtanner did not follow through on this comprehensive plan, focusing primarily on humans and the main mammals of economic significance, but others followed his lead in giving priority to genealogy. Johann Karl Illiger, for instance, endorsed a historical, genealogical conception of species in 1800, then argued for a reproductive criterion (Sloan 1979: 143).

It would be wrong to think that this historical conception of classification and species was a consensus. Many naturalists in England and Scotland either rejected or downplayed the historical, genealogical relationships, and emphasized similarity. Charles Lyell, Darwin's mentor, reported in 1832 in his *Principles of Geology* that "the majority of naturalists agree with Linnaeus in supposing that all the individuals propagated from one stock have certain distinguishing features in common which will never vary, and which have remained the same since the creation of each species" (Lyell 1997: 184–185). What is notable is, first, that Lyell was apparently unaware of Linnaeus' change of mind about the fixity of species, and, second, he was likely focusing on the views of his *British* colleagues, rather than of naturalists in general. If he had focused on the Germans naturalists who followed Kant and Buffon, he may have offered a different analysis. Lyell's own views are complicated however. He had some essentialist tendencies, but because of his geological starting point, with its tendency to think about strata historically, and the need to make sense of newly discovered fossils, his overall approach was deeply historical. This is apparent in an 1824 article in *The Quarterly Review*:

> None of these fossil plants and animals appear referable to species now in being, with the exception of a few imbedded in the most recent strata; yet they all belong to the genera, families, or orders established for the classification of living organic productions. They even supply links in the chain, without which our knowledge of the existing systems would be comparatively imperfect. It is therefore clear to demonstration, that all, at whatever distance of time created, are parts of one connected plan. (Wilson 1970: xxiii)

Two years later, he endorsed the view that the fossil discoveries revealed "that in ascending from the lowest to the more recent strata, a gradual and progressive scale could be traced from the simplest forms of organization to those more complicated, ending at length in the class of animals most related to man" (Wilson 1970: xxiv). This may not be the explicitly historical conception of species as *lineages* that we saw in Buffon, nonetheless it is not a static and unchanging conception either.

From Ray through Linnaeus and Buffon, there was a turn to a historical and genealogical conception of species. Members of a species were connected through reproduction into lineages that stretched back to the Creation. This emphasis on genealogy made similarity subservient in determining species membership. An organism was a member of a species taxon not because it was similar to other members of that species – not because they shared essential traits – but because its parents were members of that species taxon. There was nonetheless, however, a contravening practical tendency. Similarity, if based on observable, identifiable and highly variable traits, could serve to generate groupings, even if similarity was not determinative of species membership. Practical demands encouraged an emphasis on certain kinds of observable similarities. Finally, there was the tendency as well to use the Aristotelian traits associated with nutrition and reproduction – traits that were functionally "essential" to life and reproduction. The effect of all these tendencies is that, at the beginning of the nineteenth century, even though there was a powerful tendency to think of species taxa in historical, genealogical terms, there was at the same time, a variety of criteria for identifying and individuating species.

THE DARWINIAN CHALLENGE

Those who interpret Aristotle, Linnaeus and other pre-Darwinians as property essentialists usually interpret Darwin's challenge to traditional species concepts accordingly. On this view, Darwin challenged the idea that species were defined by a set of unchanging, essential properties, through his population thinking and gradualism that conceive species in terms of variability at a particular time and gradual change over time. Species taxa cannot have a set of unchanging essential properties because at any given time there will be constant variation within the members of the species, and the set of properties associated with the species will change over time. Ereshefsky explains this standard view:

> Linnaeus thought that species and other taxa are the result of divine intervention. Once a taxon is created, each of its members must have the essential properties of that taxon. The evolution of a species was foreclosed by God's original creation. Needless to say, Darwinism gives us a different picture of the organic world. Taxa are the products of natural rather than divine processes. Species are evolving lineages, not static classes of organisms. This conceptual shift in biological theory is well discussed in the literature and comes under many banners. Some authors talk of the "death of essentialism"; others refer to the "species are individuals" thesis. In broader perspective, this conceptual shift falls under "the historical turn" in biology, or what Ernst Mayr calls "population thinking." (Ereshefsky 2001: 3)

Given the preceding outline of the views of Buffon, and the many who followed him in adopting an historical conception of species, this standard view is problematic. Nonetheless there is some plausibility. Darwin not only could have, but would have agreed that species do not have a set of unchanging essential properties. And he would likely do so at least

70

partly on the basis of population thinking and gradualism. Population thinking implies variability within a species, while gradualism implies that new species are formed through a gradual change in properties, rather than discontinuous breaks that can clearly demarcate animal kinds. Nonetheless, when we look at what Darwin actually wrote in his notebooks, correspondence and published work, the Essentialism Story here is misleading. First it misleads us about what Darwin seemed to think he was doing. Second, it fails to recognize that Darwin was in many ways confronted with a species problem much like the one we find today.

Most striking of Darwin's discussion of species are the oft-quoted and analyzed statements about the unreality or arbitrariness of species. In the first edition of his *On the Origin of Species*, for instance, Darwin makes the following rather shocking claim, which seems to echo the early, seemingly nominalist claims of Buffon. "From these remarks it will be seen that I look at the term species, as one arbitrarily given for the sake of convenience to a set of individuals closely resembling each other" (Darwin 1964: 52). This is shocking in that Darwin's book is presumably about the origin of species. But if species are arbitrary groupings, and therefore presumably unreal, he has devoted an entire book to the origin of something unreal!

How we should ultimately understand this passage is the topic of the next chapter, but we can understand the background to this apparent skepticism about species. What confronted Darwin in 1859 was not the acceptance and use of a single essentialist species concept based on essential intrinsic properties. Instead he was confronted with what seems to be many ways of thinking about species. In the big species manuscript he interrupted to write the *Origin*, he listed some of these ways.

[H]ow various are the ideas, that enter into the minds of naturalists when speaking of species. With some, resemblance is the reigning idea & descent goes for little; with others descent is the infallible criterion; with others resemblance goes for almost nothing, & Creation is everything; with others sterility in crossed forms is an unfailing test, whilst with others it is regarded of no value. (Stauffer 1975: 98)

What is striking in this passage is his recognition that resemblance or similarity is only one of several ways of thinking about species. If so, then the property essentialism that is committed to conceiving species in terms of a set of unchanging essential properties could *at most* be just one of the many ways of conceiving species that confronted Darwin.

In Darwin's work he described at least six different ways of defining or conceiving species – based on morphology, fertility, sterility, geography, geology and descent. The *morphological criterion*, as described by Darwin, was applied on the basis of two factors – degree of difference and the presence of intermediate forms. In order to count as species, a form must be sufficiently distinguishable from other forms and must not be connected to other forms by a continuous chain of intermediate forms. In other words a group of individuals must be both different and distinct enough to count as a species.

> We have seen that there is no infallible criterion by which to distinguish species and well-marked varieties; and in those cases in which intermediate links have not been found between doubtful forms, naturalists are compelled to come to a determination by the amount of difference between them, judging by analogy whether or not the amount suffices to raise one or both to the rank of species. (Darwin 1964: 56–57)

On the *fertility criterion*, the ability of two individuals to successfully reproduce determines whether they are of the same species. Two individual organisms are members of different species if they cannot successfully reproduce together. And on the related *sterility criterion*, the fertility of offspring is determinative. Individuals are members of different species even if they can reproduce, but their offspring are sterile. They are of the same species if the offspring is fertile, no matter how different morphologically (Darwin 1964: 248). In his *Natural Selection*, Darwin told us (emphasis added):

> Some authors, such as Kölreuter, take the fertility of the offspring of two forms as the sole [or leading] test of what to consider as species; & *however unlike two forms may be*, if they produce quite fertile offspring, they consider them specifically as the same. (Stauffer 1975: 97)

According to the *geographic criterion*, two similar forms are more likely to be ranked as separate species if they are geographically separated: "A wide distance between the homes of two doubtful forms leads many naturalists to rank both as distinct species" (Darwin 1964: 49). On the *geologic criterion*, species determinations are made on the basis of location in the geologic strata. "It is notorious on what excessively slight differences many paleontologists have founded their species; and they do this the more readily if the specimens come from different sub-stages of the same formation" (Darwin 1964: 297).

Of all the definitions of species Darwin discussed, the most important seems to be the criterion based on genealogy or descent, and that echoes the views of Ray Linnaeus, Buffon and Kant. In fact, Darwin claimed that, for naturalists in general, it trumps all others – including one based on similarity:

> With species in a state of nature, every naturalist has in fact brought descent into his classification; for he includes in his lowest grade, or that of a species, the two sexes; and how enormously these sometimes differ in the most important characters, is known to every naturalist: scarcely a single fact can be predicated in common of the males and hermaphrodites of certain cirripedes, when adult, and yet no one dreams of separating them. The naturalist includes as one species the several larval stages of the same individual, however much they may differ from each other and from the adult ... He includes monsters; he includes varieties, not solely because they closely resemble the parent-form, but because they are descended from it. (Darwin 1964: 424)

Thomas Vernon Wollaston, a critic of Darwin, confirms this in his review of the *Origin*:

> [I]t is quite evident that there is an idea involved by naturalists in the term *species* which is altogether distinct from the fact (important though it be) of mere outward resemblance,–viz. the notion of blood-relationship acquired by all the individuals composing it, through a direct line of descent from a common ancestor. (Hull 1983: 129)

What makes an individual organism a particular kind of thing is its parentage or genealogy, not the intrinsic properties it shares with other members of the species. This is most striking, as Darwin argued relative to members of *cirripedes* or barnacles. In his eight-year-long study of barnacles, Darwin studied some barnacles that exhibited an extreme case of sexual dimorphism. On his first careful observation of these barnacles, he saw on some of them tiny black spots that appeared to be parasites. Later Darwin came to realize that these were in fact males of the species. The larger female and the much smaller male were so different that on a similarity criterion these males would certainly not be of the same species or even the same genus as the female. Darwin argued that descent – not similarity – is determinative of species membership:

> I am convinced that in the cirripede Ibla without knowledge of its descent, the male & female & its two larval stages would have formed four distinct Families in the eyes of most systematic naturalists. Again the most ill-shapen monster is rendered home to its species the instant we know its parents. (Stauffer 1975: 96)

Descent has particular significance for many of Darwin's contemporaries, as it did for those who preceded him, in part because of its compatibility with a creationist approach. Darwin explained:

> The idea of descent almost inevitably leads the mind to the first parent, & consequently to its first appearance, or creation. We see this in Morton's pithy definition of "primordial forms", adopted by Agassiz. The same idea is supreme, & resemblance goes for nothing, with those zoologists, who consider two forms, absolutely similar as far as our senses serve, when inhabiting distant countries, or distant geological times as specifically distinct. Having the idea of the first appearance of a form prominently in their minds, they argue logically that as most of the forms in the two countries or times are distinct, the distinction being in some great, in others less & less, they naturally ask, why forms apparently absolutely identical should not have been separately created, & which they in consequence would call distinct species. (Stauffer 1975: 97)

Despite the priority of the genealogical criterion, and because of all the other, different ways of thinking about species, there were substantial differences in species counts. Darwin noted one case where in a group of plants, 251 species were counted by one naturalist, and only 112 by another: "a difference of 139 doubtful forms" (Darwin 1964: 48). This is a situation much like the one that perplexes modern systematists: the use of multiple species concepts, with different implications for species grouping, and no obvious solution. Darwin responded to this problem by concluding that species were undefinable, in a letter to Hooker from December 24, 1856:

> I have just been comparing definitions of species, & stating briefly how systemic naturalists work out their subject ... It is really laughable to see what different ideas are prominent in various naturalists minds, when they speak of "species" ... in some resemblance is everything & descent of little weight – in some resemblance seems to go for nothing & Creation the reigning idea – in some descent the key – in some sterility an unfailing test, with others not worth a farthing. It all comes, I believe, from trying to define the undefinable. (Burkhardt and Smith 1990: 309)

Unfortunately, Darwin did not here explain precisely what he meant by "undefinable." As we shall see in the next chapter, there are different ways we might interpret this claim, but for now, the important thing is that Darwin thought there was as yet no adequate way of conceiving species, and of solving the species problem.

THE ESSENTIALISM STORY

I began this chapter with the orthodox Essentialism Story and its account of Linnaeus. The orthodox interpretation is that Linnaeus and many of the other pre-Darwinian naturalists followed Aristotle in classifying on the basis of division, and the possession of essential traits. Organisms are members of a species by virtue of the possession of a set of essential or necessary and sufficient properties. Consequently species are unchanging, timeless and discrete. This chapter has addressed this view, and laid out its problems. While Linnaeus accepted the fixity of species in his earliest work, soon after he allowed for the production of new species by hybridization among genera or orders. For Linneaus, species were not fixed and timeless, and whatever essences they had were genealogical, and not to be associated with a particular set of properties. Many of the other pre-Darwinian naturalists such as Ray, Buffon and Cuvier also adopted historical, genealogical or reproductive criteria, rather than one based on essential or necessary and sufficient properties.

What confronted Darwin then, was not a *property essentialism*, but a multiplicity of species concepts, based on similarity, fertility, sterility, geographic location and geologic placement and descent. And while descent could trump other criteria, placing the widely different forms of the sexes and developmental stages into singles species, it could not eliminate substantial differences in species counts. Darwin was therefore confronted with a species problem much like the one confronting us today. The obvious question to ask here is how did we get this history so wrong? Why have so many accepted the Essentialism Story? An adequate answer requires a bit of "metahistory" or a history of the history. Historian Mary Winsor has given us just this kind of account, first describing the standard reconstruction as the "essentialism story." The main ideas are outlined in the abstract of her 2006 paper, "The Creation of the Essentialism Story: An Exercise in Metahistory":

> The essentialism story is a version of the history of biological classification that was fabricated between 1953 and 1968 by Ernst Mayr, who combined contributions from Arthur Cain and David Hull with his own grudge against Plato. It portrays pre-Darwinian taxonomists as caught in the grip of an ancient philosophy called essentialism, from which they were not released until Charles Darwin's 1859 *Origin of Species*. Mayr's motive was to promote the Modern Synthesis in opposition to the typology of idealist morphologists; demonizing Plato could serve this end. Arthur Cain's picture of Linnaeus as a follower of 'Aristotelian' (scholastic) logic was

woven into the story, along with David Hull's application of Karl Popper's term, 'essentialism', which Mayr accepted in 1968 as a synonym for what he had called 'typological thinking'. (Winsor 2006a: 149)

What is most striking in the history of the Essentialism Story is that it seems to be of recent origin and based on the accounts of only a few authors, none of whom were primarily historians. Nonetheless, the views of Cain, Mayr and Hull seem to have been uncritically accepted by philosophers and biologists, even in the face of contrary evidence from primary sources, and the skepticism of various historians and philosophers (Winsor 2003: 389–90).

Both Cain and Mayr had theoretical agendas in their telling of the Essentialism Story. Mayr, for instance, was anxious to contrast the population thinking of the Modern Synthesis, with the "typological thinking" of the old, bad idealist morphology (Winsor 2006a: 152–162). Population thinking treated variation within a population as an important and crucial feature of biodiversity. The interpretation of Aristotle, Linnaeus and others as property essentialists who treated variation as unimportant or illusory helped demarcate population thinking from the views it was supposed to replace. Rhetorically, it was surely important to highlight differences. But Cain and Mayr did not just highlight differences between the old and new thinking, they seemed to have manufactured them.

Some reasons for the success of the essentialism myth are likely philosophical, perhaps in the philosophical clarity of Essentialism Story. But clearly, part of its success is due to the power of the story that it tells, with Darwin as its hero:

> In that hundred-year period, between the publication of the 10th edition of Linnaeus's *Systema Naturae* in 1758 and Darwin's (1859) publication of the *Origin of Species*, natural history was progressing dramatically, with museum and herbarium collections growing, workers better trained, and the number of named taxa at all levels increasing at an explosive rate. The notion that botanists and zoologists, during this busy period of achievement, were frozen in the grip of ideas derived from Plato, Aristotle and the medieval scholasticism makes Darwin's breakthrough nearly miraculous. The story of the dominance of essentialism is as dramatic, in its way, as the myth that has Columbus's crew fearing they would fall off the edge of the earth, and I believe it is equally fictitious. (Winsor 2003: 388)

The story of Darwin battling essentialism – and winning – surely has appeal to the modern biologists and philosophers who perennially celebrate Darwin day. I do not wish to detract from the truly remarkable contributions of Darwin, but, as we shall see in the next chapter, his contributions are harder to understand and describe than the Essentialism Story suggests. The true story is much more complicated, and in many ways more interesting.

4

Darwin and the proliferation of species concepts

The species problem we have today looks a lot like the problem that confronted Darwin, but with one major difference. The modern problem is informed by Darwin's theory of evolution and recent developments of that theory. Yet it is nonetheless difficult to pin down precisely what Darwin directly contributed to the species debate. In part this is because it is difficult to pin down Darwin's views about the nature of species. In his published work, notebooks and correspondence, he wrote many seemingly inconsistent things about species. Consequently there is little consensus among historians and philosophers about his views. Some read Darwin as a species nominalist, believing that there are no species, only individual organisms. Species taxa are then nothing more than convenient ways of grouping organisms, and do not represent real things in nature. Others similarly read Darwin as a conventionalist, believing that species are just what competent naturalists identify as species. Yet others read him as accepting the reality of species taxa, but denying the reality of the species category. Still others read him as a full-blooded realist, accepting the reality of both species taxa and the species category.

This conflict in interpretation is largely due to a genuine paradox in Darwin's views. The title and contents of his most famous book, *On Origin of Species*, seem to presuppose the reality of species. Here we find discussions of variation within species, the development of new species, the extinction of species and the processes that are relevant to the identification and individuation of species. Each of these discussions seems to presuppose that species taxa exist. Yet Darwin also made many pronouncements over his long career that seemed to deny the reality of species. In this chapter, I will first address this paradox, and show how we might better understand Darwin's views based on his distinction between species and varieties. Second, I will look at the thinking about species that followed, from the Mendelians and Biometricians to the Modern Synthesis and more contemporary approaches. Here I will sketch out the many species concepts that are currently being debated in the literature. This will prepare us to understand the basis for the solution to the species problem presented in the next chapter.

In his correspondence, notebooks and published works, Darwin seemed to cast doubt on the idea that species are real things, claiming that they are nothing more than convenient yet arbitrary ways to group organisms. In an 1837 letter to Charles Lyell, after a discussion of the birds of the Galapagos he lamented: "I have been attending a very little to species of birds, & the passage of forms, do appear frightful – everything is arbitrary" (Burkhardt and Smith 1987: July 30, 1837). Nearly two decades later, in his 1856 *Natural Selection*, the long manuscript that served as the basis for the *Origin*, he observed:

> It seems to me that the term species is one arbit[r]arily given for convenience sake to a set of individuals closely like each other; &, that it is not essentially different from the term variety, which is given to less distinct & more fluctuating form. (Stauffer 1975: 166)

Three years later, in his *Origin of Species,* he was similarly skeptical at the beginning of his chapter on variation in nature: "From these remarks it will be seen that I look at the term species, as one arbitrarily given for the sake of convenience to a set of individuals closely resembling each other" (Darwin 1964: 52). And at the end of the *Origin* he made a similar claim: "We shall have to treat species in the same manner as those naturalists treat genera, who admit that genera are merely artificial combination made for convenience" (Darwin 1964: 485). It is difficult to read these passages as anything other than a denial that species are real things in nature.

Darwin also sometimes seemed to advocate a conventionalism: species are just what we conventionally identify as species, and nothing more. For instance in his *Natural Selection*:

> In the following pages I mean by species, those collections of individuals, which have commonly been so designated by naturalists. Everyone loosely understands what is meant when one speaks of the cabbage, Radish & sea-kale as species; or of the Broccoli, & cauliflowers as varieties. (Stauffer 1975: 98)

And in the *Origin* a similar view: "[I]n determining whether a form should be ranked as a species or a variety, the opinion of naturalists having sound judgment and wide experience seems the only guide to follow" (Darwin 1964: 47). While Darwin does not elaborate on this point, if species are *nothing more* than what naturalists designate at species, then again it is difficult to see how they can have a reality independent of what naturalists think.

But then in other places, Darwin seems instead to have been committed just to the view that species cannot be defined – without necessarily denying their reality. He did this in the first sentences of his chapter on "Variation under Nature" in *Natural Selection*:

> In this Chapter we have to discuss the variability of species in a state of nature. The first & obvious thing to do would be to give a clear & simple definition of what is meant by a species; but this has been found hopelessly difficult by naturalists, if we may judge by scarcely two having given the same. (Stauffer 1975: 95)

In the *Origin* Darwin lamented the lack of a satisfactory definition, then wrote:

> It must be admitted that many forms, considered by highly-competent judges as varieties, have so perfectly the character of species that they are ranked by other highly competent judges as good and true species. But to discuss whether they are rightly called species or varieties before any definition of these terms has been generally accepted, is to vainly beat the air. (Darwin 1964: 49)

Unfortunately, Darwin did not elaborate in these passages, so we do not know precisely what he had in mind. He might have just been worried that we have no definition of species and need to get one. Or it may be that he thought there were different kinds of species, so no *single* definition will do. This pluralistic worry suggests that we have no single species definition and cannot have one, because there is no *single* kind of species thing. Notice this pluralism need not imply the antirealism we find in Darwin's words quoted above. Different kinds of species might all be real in straightforward, albeit different, ways. Notice also the ability to give a definition of a thing does not imply its reality. Just as we can give definitions of imaginary numbers and impossible geometric figures, we can give a definition of the *species* term – whether or not there are any species. We could simply define species as "the things that naturalists believe to be species," or as "arbitrary groupings for convenience."

On the other hand, Darwin sometimes wrote as if species taxa were real and well defined. In the chapter "Difficulties on Theory" in the *Origin*, he wrote: "To sum up, I believe that species come to be tolerably well-defined objects, and do not at any one period present an inextricable chaos of intermediate links" (Darwin 1964: 177). And near the end of the *Origin*, Darwin actually seemed to give a definition of the species category: "On the view that species are only strongly marked and permanent varieties, and that each species first existed as a variety, we can see why it is that no line of demarcation can be drawn between species" (Darwin 1964: 469). These statements seem more consistent with the implicit assumption of realism in much of his discussion in the *Origin of Species*. Time after time we find Darwin arguing as if various species taxa are real, and that the category *species* referred to real things in the world. When he wrote, for instance, that "It need not be supposed that all varieties or incipient species necessarily attain the rank of species" (Darwin 1964: 52), surely he was implying that both varieties and species are real things.

INTERPRETING DARWIN

The interpretations of Darwin's statements about species are as conflicted as his views. This is not surprising. One commentator could focus on Darwin's claims that species groupings were arbitrary, another could focus on his claims that it was mere convention, and yet another could focus on his presuppositions of species reality. Ernst Mayr is one who focused on the apparently antirealist passages, quoting Darwin directly:

> In Darwin, as the idea of evolution became firmly fixed in his mind, so grew his conviction that this should make it impossible to delimit species. He finally regarded species as something purely arbitrary and subjective. "I look at the term species as one arbitrarily given for the sake of convenience to a set of individuals closely resembling each other..." (Mayr 1957a: 4)

According to Mayr, this nominalism and antirealism was adopted by the Darwinians who followed:

> The seventy-five years following the publication of the *Origin of Species* (1859) saw biologists rather clearly divided into two camps, which we might call, in a somewhat oversimplified manner, the followers of Darwin

81

and those of Linnaeus. The followers of Darwin, which included the plant breeders, geneticists, and other experimental biologists, minimized the "reality" of objectivity of species and considered individuals to be the essential units of evolution. Characteristic for this frame of mind is a symposium held in the early Mendelian days, which endorsed unanimously the supremacy of the individual and the nonexistence of species. Statements made at this symposium ... include the following: "Nature produces individuals and nothing more ... Species have no actual existence in nature. They are mental concepts and nothing more ... Species have been invented in order that we may refer to great numbers of individuals collectively. (Mayr 1957a: 4)

David Hull explains the argument that underlies this antirealism, as it is attributed to Darwin, as based on a gradualism:

The only basis for a natural classification is evolutionary theory, but according to evolutionary theory, species developed gradually, changing one into another. If species evolved so gradually, they cannot be delimited by means of a single property or set of properties. If species can't be so delimited, then species names can't be defined in a classic manner, then they can't be defined at all. If they can't be defined at all, then species can't be real. If species aren't real, then "species" has no reference and classification is completely arbitrary. (Hull 1992a: 203)

There are two things to note here. First this assumes that Darwin was responding to a property essentialism in his thinking about species. As I argued in chapter 3, this is a doubtful assumption. Darwin was responding to a variety of ways of thinking about species, including what he thought was the primary way – in terms of genealogy. A passage from the *Origin*, and quoted in chapter 3 is worth reproducing here:

With species in a state of nature, every naturalist has in fact brought descent into his classification; for he includes in his lowest grade, or that of a species, the two sexes; and how enormously these sometimes differ in the most important characters, is known to every naturalist: scarcely a single fact can be predicated in common of the males and hermaphrodites of certain cirripedes, when adult, and yet no one dreams of separating them. The naturalist includes as one species the several larval stages of the same individual, however much they may differ from each other and from the adult ... He includes monsters; he includes varieties, not solely because they closely resemble the parent-form, but because they are descended from it. (Darwin 1964: 424)

The second thing to note is that this antirealist interpretation seems to ignore all those instances where Darwin wrote as if individual species taxa were real and that the species category had theoretical significance. Nonetheless, biologists from Stephen Jay Gould and Niles Eldredge to the philosophers Elliott Sober, David Hull and Marc Ereshefsky can be understood as endorsing similar antirealist interpretations of Darwin's views (Stamos 2007: 1–20).

Perhaps we cannot interpret Darwin so that everything he said about species is consistent. The most reasonable interpretation *may be* that Darwin was simply confused or unclear about some of the philosophical issues. There is ample reason to arrive at this conclusion. In his discussions of species, Darwin did not always clearly distinguish the species *category* from various species *taxa*. In a passage quoted above, for instance, he tells us that: "I was much struck how entirely arbitrary the distinction is between varieties & species, when I witnessed different naturalists comparing the organic productions which I brought home from the islands, off the coast of S. America" (Stauffer 1975: 115). Here it could be the distinction between the categories *species* and *varieties* that is at issue, or it could be the distinction between particular groups of organisms that get identified as species or varieties.

Nor was Darwin clear in his use of the term *definition*. Sometimes it seemed to refer to the meaning of a term, other times to lines of demarcation in reference. In Darwin's usage, the definition of the species *category* might be either an account of what a *term* such as *species* means, or it might be an account of how to distinguish species kinds from varieties kinds. Similarly, the definition of a species *taxon* could be a list of properties that are associated with that taxon, or it could be a way of demarcating members of that species taxon from members of other species taxa. These are not equivalent uses of *definition* and its cognates. One can say what distinguishes birds from reptiles without giving anything like a full definition of either taxon. And one can say what distinguishes species from varieties without giving anything like of full definition of either category. This in fact seems to be what Darwin was doing when he suggested that "species are only strongly marked and permanent varieties" in the passage quoted above. Conversely, one might be able to say what the meaning of a term is without being able to draw clear and unambiguous lines of demarcation. The lines of demarcation may be arbitrary in the sense that we could reasonably draw them in a number of different ways, without the meaning of the terms being arbitrary.

These apparent inconsistencies in Darwin's writings have not dissuaded commentators from trying to interpret him in a clear, consistent way. One standard interpretation is that Darwin was a realist relative to species taxa, but not relative to the species category. Michael Ghiselin, John Beatty and Marjorie Grene and David Depew (Grene and Depew 2004: 212–213) seem to interpret Darwin in this way. Ghiselin for instance, tells us that "in technical terms, Darwin was denying the reality, not of taxa, but of categories" (Ghiselin 1969: 93). Beatty credits Ghiselin for this view, then seems to endorse it:

> Darwin denied that there was a definition of "species" that excluded all of what were called "varieties", or a definition of "variety" that excluded all of what were called "species". But he affirmed the reality of what naturalists called "species" and of what they called "varieties" – of what were given species and variety names. (Beatty 1992: 240)

But this is all puzzling. Beatty goes on to tell us that for Darwin species were "chunks within the genealogical nexus of life" (Beatty 1992: 240). It is not clear why Darwin could not see this as at least part of a definition. Darwin does, in fact, give something like a definition in the passage quoted above, that "species are only strongly marked and permanent varieties" (Darwin 1964: 469). What really seems to be the issue with Beatty (and perhaps others) is whether there is a definition that can unambiguously demarcate *all species things from all variety things*. According to Beatty, the terms *species* and *variety:*

> did not refer to *one kind* of chunk with their species names and to *another kind* of chunk with their variety names. That was why there was no definition of "species" that excluded all of what were called "varieties", and so on. Nevertheless, their names referred to real genealogical segments in each case. (Beatty 1992: 240)

But surely it is too much to require that a definition be completely determinative and unambiguous in its application. This requirement would seem to conflate the two senses of *define* – as providing a guide to meaning, or as to providing lines of demarcation in reference. It would also make it impossible to define terms like *village, town* and *city* that have vague boundaries of application, but nonetheless are meaningful.

Perhaps we can avoid this muddle with the third position, that Darwin accepted the reality of both the species category and species taxa. This is the view advocated by David Stamos, who first argues for the distinction between horizontal entities and vertical entities. The idea is that we can conceive of groups of organisms either vertically in terms of an extended period of time (diachronically), or horizontally in terms of a population at a single, minimal time frame (synchronically) (Stamos 2007: 37). According to Stamos, what made the species category and species taxa real for Darwin was the existence of a set of "objectively and universally valid criteria" for identifying horizontal species – the morphological distinctness and constancy produced by natural selection (Stamos 2007: 86–88).

> [I]n Darwin's view a species is a primarily horizontal similarity complex of organisms morphologies and instincts distinguished at any horizontal level primarily by relatively constant and distinct characters of adaptive importance from the viewpoint of natural selection. (Stamos 2007: 127)

Stamos argues that the reality of both the species category and species taxa is based on the role of natural laws in the formation and maintenance of species, and in particular, the law of natural selection. Among these laws are those that assert the necessity of evolution, the improbability of evolution reversing or repeating itself, the inevitability of extinction and more (Stamos 2007: 32–33). So, according to Stamos, the species category is real for Darwin because it contains things subject to scientific laws, with objective and universal criteria. And species taxa are real because they are instances of the species category – they satisfy these objective criteria.

If Stamos were right about Darwin's views, we would still have to explain all those passages in Darwin's writings where he seems to deny both the reality of species and the possibility of defining species. Stamos, following the example of Beatty, proposes a "strategy theory" to render Darwin's statements consistent with his alleged species category realism. Beatty had argued that Darwin recognized that he must use the language of his contemporaries about species, if he were to engage them in discussion. This required that he accept the standard usage of the term *species*. But he also rejected the way his contemporaries conceived of species. So when he seemed to be an antirealist about species, he was rejecting the *meaning* of the term *species* as his contemporaries used it, even though he still accepted the *application* of that term.

> Darwin's theory of the evolution of species was undermined by the non-mutationist and non-transmutationist definitions of "species" to which his fellow naturalists adhered. He clearly could not defend the evolution of species, in any of those senses of "species". He could and did defend, however, the evolution of what his fellow naturalists actually called "species" – on the supposition that what they called "species" did not satisfy their non-evolutionary definition of "species". (Beatty 1992: 242)

On Stamos' variant of the strategy theory, Darwin was doing two things. First, he was giving a *reductio ad absurdum* argument that if varieties were not real, and species were like varieties, then species were not real either. This contradicted the assumption of Darwin's contemporaries that species were real in a non-evolutionary, essential way (Stamos 2007: 167–8). Second, Darwin would speak the language of non-realism with skeptics, until they became true believers. Then he would shift to the language of realism (Stamos 2007: 170–186). The advantage of this is primarily rhetorical. Darwin could reject the reality of species on a meaning of the term *species* that he opposed, without denying the reality of species in other more acceptable senses.

One worry about this strategy theory, as should be clear from the last chapter, is its apparent assumption about what Darwin's contemporaries might mean by the term *species*. It would make sense if his contemporaries accepted a property essentialism that conceived species in terms of unchanging essential properties. Darwin surely would think that species were not real in this sense. But there was no consensus among Darwin's contemporaries about the meaning of the term species. Some of his contemporaries identified species with lineages. Others with interbreeding groups. It is not so clear that he would be antirealist about species in this sense.

SPECIES AND VARIETIES

Whether or not any of the strategy theories succeeds in making Darwin's views consistent is not of primary interest here. It may be just that Darwin was confused – as many still are – about how to best conceive species. Present purposes are better served by looking more closely at a continuing theme in Darwin's writings – the differences between species and varieties. When Darwin gave us his apparently antirealist views, typically they were framed in terms of the distinction between species and varieties. This distinction is explicit in a passage from his *Natural Selection*. "I was much struck how entirely arbitrary the distinction is between varieties & species" (Stauffer 1975: 115). Significantly, Darwin also formulated many of his apparently realist claims in terms of the species–varieties distinction:

> On the view that species are only strongly marked and permanent varieties, and that each species first existed as a variety, we can see why it is that no line of demarcation can be drawn between species. (Darwin 1964: 469)

We can, I believe, make some progress in understanding the species problem as Darwin saw it – by consideration of how some varieties come to be species by becoming "strongly marked and permanent."

The first thing to notice is that, for Darwin, there is no simple dichotomy between varieties and species. Rather there is a continuum from individual variations to varieties, subspecies and legitimate species, as he explained in his *Origin*:

> Certainly no clear line of demarcation has as yet been drawn between species and sub-species – that is, the forms in which the opinion of some naturalists come very near to, but do not quite arrive at the rank of species; or, again, between sub-species and well-marked varieties, or between lesser varieties and individual differences. These differences blend into each other in an insensible series; and a series impresses the mind with the idea of an actual passage. (Darwin 1964: 51)

This suggests that varieties, subspecies and species are not so much different in kind, as different in degree. This is perhaps what Darwin had in mind when he claimed, in a passage quoted above, that the term species "is not essentially different from the term variety, which is given to less distinct and more fluctuating form" (Stauffer 1975: 166).

The second thing to notice is that we can think about this continuum in terms of the empirical, observable similarities and differences we see in a series representing a group of organisms at a single time. Or alternatively, we can think about the continuum in terms of an "actual passage" – the change in a series of organisms over the passage of time. In other words we can think of the distinction between species and varieties *synchronically* – at a single time – or *diachronically* – over time. A naturalist is thinking synchronically when considering a group of organisms existing at roughly a single time – as Darwin did in his study of barnacles or finches. A paleontologist, on the other hand, is typically thinking diachronically, when considering the change in a group of organisms over significant time scales. And an evolutionist such as Darwin can also think diachronically about the development of a variety from individual differences, and the transformation of a variety into a species. Immediately following the passage from the *Origin* just quoted, he asked us to think about the species–variety distinction in these two ways:

> I look at individual differences, though of small interest to the systematist, as of high importance for us, as being the first step towards such slight varieties as are barely thought worth recording in works on natural history. And I look at varieties which are in any degree more distinct

and permanent, as steps leading to more strongly marked and permanent
varieties; and at these latter, as leading to sub-species, and to species ...
Hence I believe a well-marked variety may justly be called an incipient
species. (Darwin 1964: 51–52)

It is surely not insignificant that this passage just preceded one most
often quoted as antirealist:

> From these remarks it will be seen that I look at the term species, as one
> arbitrarily given for the sake of convenience to a set of individuals closely
> resembling each other, and that it does not essentially differ from the term
> variety, which is given to less distinct and more fluctuating forms. The
> term variety, again, in comparison with mere individual differences, is also
> applied arbitrarily, and for mere convenience sake." (Darwin 1964: 52)

These passages, all taken together, seem to imply the following. First,
there is a continuum from individual variations to varieties and species.
Second, we can observe this continuum at a single time, when we look at
the various coexisting series of differences in organisms. But we can also
think about it over time, as individual differences sometimes become
varieties, that in turn sometimes become species. Third, what seemed
arbitrary to Darwin may be just where we draw lines in this continuum.
It is not that the differences between varieties and species are not real!
Finally, this demarcation problem is both practical and theoretical. It is
often not clear where to draw the line in practice when grouping organisms into species. Nor is it clear how to formulate the criteria used to
draw the lines between varieties and species. As we shall see, the criteria, as understood by Darwin, were typically vague and formulated in
terms of "more or less."

The two ways of thinking about the continuum – synchronically and
diachronically – imply two distinct questions: first, how should one distinguish species and varieties at a particular time; second, how should
one distinguish species and varieties over time, as a variety becomes
a species? While no single answer may be adequate to both questions,
nonetheless there is a set of criteria that is potentially relevant to both.
Understanding how Darwin saw this set of criteria is complicated in
three ways. First, he seemed to think that the criteria were often applied
in appropriate ways by his fellow naturalists – even though they were
not typically evolutionists. Second, which criterion was most important
depended on context – the organism in question and the relevant evolutionary processes. Third, the synchronic context was observable in a way
the diachronic was not, but the synchronic context also depended on the

diachronic. Darwin thought that whether a difference is relevant to the species–variety distinction depended on its implication for the history – the evolutionary fate – of the group of organisms. This last point will be obvious as we look more closely at Darwin's criteria for distinguishing species and varieties, based on morphology, descent, fertility and sterility, and geography.

In a passage quoted above, Darwin seems to have suggested a morphological criterion when he tells us that "more strongly marked and permanent varieties" ultimately lead to subspecies and species. (Darwin 1964: 51–52) Implicit here are three main morphological criteria. The first is based on the idea that species are more distinct than varieties in the relative absence of intermediate forms. We might formulate it as follows:

Morphological distinctness criterion: *the members of different species can be distinguished from those of varieties* **to the degree** *they are morphologically distinct – have gaps in the morphological series that unite them.*

In general, if there is an uninterrupted series of variations, then the organisms are to be grouped into a set of varieties of a species. Sometimes one of the varieties is seen as the predominant form – the species form – depending on priority or number of members. Darwin described this practice:

> Practically, when a naturalist can unite two forms together by others having intermediate characters, he treats the one as a variety of the other, ranking the most common, but sometimes the one first described, as the species and the other as the variety. (Darwin 1964: 47)

If there are gaps in the morphological series, and to the degree there are gaps, the groupings should be conceived as different species rather than mere varieties.

The second morphological criterion is based on the idea that there is a greater difference between the forms of species than the forms of varieties. We might formulate this as follows:

Morphological difference criterion: *the members of different species can be distinguished from those of varieties* **to the degree** *that there is a greater morphological difference.*

In support of this criterion, Darwin again cited the actual practice of naturalists:

> We have seen that there is no infallible criterion by which to distinguish species and well-marked varieties; and in those cases in which intermediate

links have not been found between doubtful forms, naturalists are com-
pelled to come to a determination by the amount of difference between
them, judging by analogy whether or not the amount suffices to raise one
or both to the rank of species. Hence the amount of difference is one very
important criterion in settling whether two forms should be ranked as
species or varieties. (Darwin 1964: 56–57)

For Darwin the differences must also be permanent. Fluctuating differences due to environment are not as significant as those that are stable.
A tree that is stunted in growth due to its high altitude is not therefore of
a different species than one of greater size at a lower altitude. Similarly,
differences that are not inherited are not significant for species designations. The third morphological criterion is permanence or constancy of
character:

Morphological constancy criterion: *the members of different species
can be distinguished from those of different varieties* **to the degree** *that
their characters are more permanent and constant.*

So even if there is little difference between two forms, if the differences are constant and permanent, then they could be indicative of good
species.

> Those forms which possess in some considerable degree the character
> of species, but which are so closely similar to some other forms, or are
> so closely linked to them by intermediate gradations, that naturalists do
> not like to rank them as distinct species, are in several respects the most
> important for us. We have every reason to believe that many of these
> doubtful and closely-allied forms have permanently retained their char-
> acters in their own country for a long time; for as long, as far as we know,
> as we have good and true species. (Darwin 1964: 47)

Notice that individually these criteria are all a matter of more or less.
The gaps between forms are always more or less; the amount of difference
is always more or less; and the constancy of characters is always more or
less. So there are no sharp lines to draw between species and varieties on
the basis of any *single* criterion – even if there are unproblematic cases
of good gaps, large differences and highly constant characters. But also,
since there are three criteria, there is always the problem of weighting.
Which criterion should count the most, and how much more than the
others? One naturalist could emphasize the size of the gap, another the
amount of difference, and third the constancy of differences. And even
if these are relatively clear-cut cases with respect to each criterion, there
could still be vagueness in weighting criteria. A group of organisms may

be connected to others by many intermediate links, but still have a large amount of permanent difference. In this case, the naturalist would simply have to choose which criterion to apply.

But there is another issue lurking here. For Darwin and the naturalists who preceded him, morphology was important in species groupings partly because it was observable and therefore practical. But for Darwin it was also important because morphological difference was sometimes a sign of the divergence associated with evolution. The only diagram appearing in the first edition of the *Origin* was a tree that represented the branching and splitting of lineages. The process of splitting and divergence represented in this tree is the diachronic aspect of species. It is not just that species are generally more distinct, different and permanent than varieties, but they are the result of evolutionary processes that make them so, and that are ultimately responsible for continuing splitting and morphological divergence into higher-level taxa such as genera, classes and orders. Darwin makes this clear in his discussion of his principle of the divergence of character:

> The principle, which I have designated by this term, is of high importance on my theory, and explains, as I believe, several important facts. In the first place, varieties, even strongly-marked ones, though having somewhat the character of species – as is shown by the hopeless doubts in many cases how to rank them – yet certainly differ from each other far less than do good and distinct species. Nevertheless, according to my view, varieties are species in the process of formation, or are, as I have called them, incipient species. How, then, does the lesser difference between varieties become augmented into the greater difference between species? (Darwin 1964: 111)

Here Darwin was asking us to consider the processes that operate in making incipient varieties into species. These processes can potentially provide criteria for distinguishing varieties from species.

The most obvious process whereby varieties become species is the operation of natural selection. Any particular geographic area can support more life if organisms diversify and vary in habits of life by diverging from the ancestral form. Intermediate forms will tend to be eliminated because they will be competing with both sets of divergent forms, while the *most* divergent forms will have the least competition. A form that is intermediate in size, for instance, will likely be in competition with both its larger and smaller cousins in terms of food and shelter, while the largest and smallest forms will only be in competition with the intermediate form. This divergence, Darwin claims, leads to a branching process in evolution as illustrated by the tree diagram.

> I request the reader turn to the diagram illustrating the action, as formerly explained, of these several principles; and he will see that the inevitable result is that the modified descendants proceeding from one progenitor become broken up into groups subordinate to groups. (Darwin 1964: 412)

It is in this branching and divergence that varieties become species. Varieties become increasingly distinct from other varieties as intermediate forms are eliminated, and as they become increasingly different in terms of morphology. As these differences are created and maintained by natural selection, they are also relatively constant and permanent. Thus, for Darwin, natural selection explains and justifies the morphological criteria that naturalists have tended to use to distinguish varieties and species. Consequently, the judgment of competent naturalists, *insofar as they use these three criteria*, was a reasonable guide to species identifications.

This principle of divergence can help us understand other criteria employed by Darwin to distinguish species from varieties: fertility and sterility, and geography. As noted in the previous chapter, many of Darwin's predecessor's regarded fertility and sterility as species criteria. If two individuals can interbreed and produce fertile offspring, they were typically regarded as members of the same species. If they could not reproduce fertile offspring, then they were taken to be members of different species. Darwin explained this at the beginning of his chapter on sterility in the *Origin*: "The view generally entertained by naturalists is that species, when intercrossed, have been specially endowed with the quality of sterility, in order to prevent the confusion of all organic form" (Darwin 1964: 245). Darwin seemed to endorse this criterion for species:

> The fertility of varieties, that is of the forms known or believed to have descended from common parents, when intercrossed, and likewise the fertility of their mongrel offspring, is, on my theory, of equal importance with the sterility of species; for it seems to make a broad and clear distinction between varieties and species. (Darwin 1964: 246)

The idea here is fairly simple. If fertility is maintained among members of different varieties, then these varieties will be unlikely to continue diverging into the different forms associated with species. Interbreeding would instead produce a blending of forms. So part of the process of divergence is the reduction of fertility and development of sterility among those of different and diverging groups. As might be expected from the assumption that sterility itself evolves, Darwin believed fertility and sterility to be a matter of degree.

> It is certain, on the one hand, that the sterility of various species when crossed is so different in degree and graduates away so insensibly, and, on the other hand, that fertility of pure species is so easily affected by various circumstances, that for all practical purposes it is most difficult to say where perfect fertility ends and sterility begins. (Darwin 1964: 248)

We might formulate this criterion as follows:

Fertility/sterility criterion: *the members of different species can be distinguished from those of different varieties **to the degree** that either they cannot successfully reproduce, or that their offspring is sterile.*

Here, as in the morphological criteria, there is a rule for grouping organisms into species and varieties, even though it is a matter of degree, and the application of that rule is therefore sometimes uncertain.

Darwin accepted the morphological and fertility/sterility criteria employed by his fellow naturalists, even though they employed them in the absence of his evolutionary framework. He also followed their example in employing a geographic criterion, albeit not in precisely the same way. Typically two forms that were geographically distinct were ranked as different species, even if they were relatively similar morphologically. Like the other criteria, this one is problematic because it is a matter of degree:

> A wide distance between the homes of two doubtful forms leads many naturalists to rank both as distinct species; but what distance, it has been well asked will suffice? if that between America and Europe is ample, will that between the Continent and the Azores, or Madeira, or the Canaries, or Ireland be sufficient? (Darwin 1964: 49)

But for Darwin, it was not the geographic separation that was most important, but geographic *range*, as he explained in his *Natural Selection*:

> When two varieties inhabit two distinct countries, as is often the case & as is very generally the case with the higher animals, it is obvious that the two varieties separately have a much narrower range than the parent species. A variety, for instance, inhabiting N. America & another variety of the same species inhabiting Europe will both have a very much more confined range than the parent form, so on a much smaller scale, the many varieties of endemic species, confined to the separate islets of the same

small archipelago ... follow the same rule ... These considerations alone make it probable that the far greater number of varieties have narrower ranges than the species whence they have sprung. (Stauffer 1975: 138)

In his *Origin* he similarly endorsed this principle:

Varieties generally have much restricted ranges: this statement is indeed scarcely more than a truism, for if a variety were found to have a wider range than that of its supposed parent-species, their denominations ought to be reversed. But there is also reason to believe, that those species which are very closely allied to other species, and in so far resemble varieties, often have much restricted ranges. (Darwin 1964: 58)

The last sentence here suggests this criterion is a matter of degree, with good and distinct species having proportionally greater range, and with marginal species and mere varieties each having proportionally smaller ranges. If so, we might formulate the criterion as follows:

*Geographic criterion: the members of different species can be distinguished from those of different varieties **to the degree** that they are less localized into a geographic area.*

Here, as in the previous criteria, there is a continuum and both the criterion and the distinction between species and varieties is one of degree.

The reasoning behind this criterion, like the other criteria, is based on the principle of divergence, which is itself based on geographic localization.

The truth of the principle, that the greatest amount of life can be supported by great diversification of structure, is seen under many natural circumstances. In an extremely small area, especially if freely open to immigration, and where the contest between individual and individual must be severe, we always find great diversity in its habitants. (Darwin 1964: 114)

In a restricted geographic area, natural selection can be most severe and eliminate intermediate forms, thereby producing the divergence from varieties to species. Intermediate forms, since they will typically be subject to greater competition, will be more likely to be eliminated. As the remaining forms continue to diverge, they gradually lose the ability to reproduce, becoming good and distinct species. Then these distinct species will spread and the process will repeat in various isolated geographic areas. New varieties develop from individual variations that then become species. Here, as with the other criteria, the distinction

94

will be a matter of degree. Good and distinct species will typically have wider ranges than marginal species, which will in turn have wider ranges than varieties, that obviously have wider ranges than mere individual variations.

There are three things to notice about Darwin's account of the criteria for distinguishing species and varieties. First, the criteria are primarily *practical* rules for drawing the lines between species and varieties in the grouping and ranking of organisms. Given a group of organisms that share some morphological similarities, but are different in other ways, how should they be grouped and ranked? Should these groupings be considered mere varieties, true species or something in between? This is not to say, however, that these criteria are merely practical and can therefore be applied independently of what we might regard as theoretical accounts of the species and variety categories. For Darwin, the theory of evolution by natural selection has implications for what constitutes species and varieties, and how they should be distinguished. Repeatedly, he referred to the principle of divergence to give guidance. This principle, in Darwin's view, could tell us how to distinguish species and varieties on the basis of morphology, fertility/sterility and geographic range.

Second, Darwin thought that competent naturalists were often already using good criteria in classifying organisms into varieties and species, even if they did not do so clearly, unequivocally and on the basis of an evolutionary framework. They often got the practical grouping principles right, even if they did not have the correct theoretical basis. This is why he could say without contradicting himself that "in determining whether a form should be ranked as a species or a variety, the opinion of naturalists having sound judgment and wide experience seems the only guide to follow" (Darwin 1964: 47). This is not to say, however, that the opinion of naturalists with sound judgment could not be corrected by consideration of the principle of divergence and how it generates new species from mere varieties. Darwin did this repeatedly, arguing how each criterion could be better understood through the principle of divergence to draw the lines between species and varieties better. Relative to the geographic criterion, for instance, he argued that it was not the distance between geographic ranges, but the size of the geographic range that was most relevant.

The third thing to notice is that the application of these criteria is equivocal both individually and collectively. First, each criterion is gradual and has what philosophers call borderline vagueness. Morphological distinctness, difference and permanence all come in degrees, as does fertility/sterility and geographic localization. So one naturalist could reasonably draw the line in one place between species and variety, while another could reasonably draw a line elsewhere. But perhaps even more importantly, since there are multiple criteria, the choice and weighting of criteria can also be done differently by naturalists and with respect to different organisms. One naturalist can employ morphological criteria exclusively. Another could look at fertility and sterility. Or a naturalist could use one criterion with one group of organisms, and a different one with another group. This implies that, even though the criteria are all legitimate and useful, there will always be an element of arbitrariness. So Darwin's statements about the arbitrariness of species designations, need not be taken as rejecting the real value of the criteria for distinguishing species, and the significance of the grouping and ranking of organisms into species and varieties. In other words, that there is an arbitrariness does not imply that there are no such things as species taxa distinguishable from variety taxa. Nor does it imply that the species category, as distinguished from the variety category, is itself unimportant or arbitrary.

Perhaps what is most important here is that *in the absence of evolutionary theory*, there may be no way to decide which criterion is most relevant in a particular case, and where to draw the lines relative to each criterion. But on the assumption of evolution, we can give reasons to apply a single criterion in a particular way and to prefer one criterion over another. In sexual organisms, for instance, reproduction can blend differences and prevent divergence. So a fertility/sterility criterion needs to be taken seriously. Darwin repeatedly appealed to his principle of divergence to understand how the various criteria based on morphology, fertility/sterility and geography can legitimately be used to distinguish species and varieties. Those who followed Darwin similarly turned to evolutionary theory for guidance on how to group organisms, and rank them into species and varieties.

THE MODERN SYNTHESIS

After Darwin we find the development of more specialized fields in biology: embryology, cytology, genetics, behavioral biology, ecology and more. These newer fields aspired to be experimental, were often modeled on the physical sciences, and typically less dependent on systematics than the zoology and botany of Darwin's time. For many of the evolutionists in these newer fields, the emphasis was on the development of organisms and evolutionary transformation rather than biological diversity and the proliferation of species taxa. While there may have been less interest in the species problem and competing species definitions from the last part of the nineteenth through the first part of the twentieth century, the worries and the theorizing did not entirely cease.

At the end of the nineteenth century there was a heated debate about the nature of variation and speciation. The "saltationists," Francis Galton, William Bateson and Hugo de Vries, argued that the variation that led to new species was discontinuous, made up of "saltations" or "jumps" in form. New species were the product of discrete changes in form, rather than gradual change built up by natural selection. After the rediscovery of Mendel's laws at the turn of the century, these saltations were identified with mutations. Hugo de Vries, in his 1905 *Species and Varieties,* asked: "What are species? Species are considered as the true units of nature by the vast majority of biologists" (de Vries 1905: 32). But then he questioned how species designations are typically made: "Genera and species are, at the present time, for a large part artificial, or stated more correctly, conventional groups. Every systematist is free to delimit them in a wider or a narrower sense, according to his judgment" (de Vries 1905: 36). He then lamented the wide disparity in species counts:

> In the Handbook of the British Flora, Bentham and Hooker describe the forms of brambles under 5 species, while Babbington in his Manual of British Botany makes 45 species out of the same material. So also in other cases; for instance, the willows which have 13 species in one and 31 species in the other of these manuals, and the hawkweeds for which the figures are 7 and 32 respectively. (de Vries 1905: 36–7)

De Vries does so, in part, as an argument for his own conception of "elemental species," conceived as part of his "mutation" theory. Species, for de Vries were to be identified and distinguished by the appearance of new mutations – new, distinct unit characters. "True elementary species differ from each other by unit-characters. They have arisen by progressive mutation. Such characters are not contrasting. One species has one kind of unit, another species has another kind" (de Vries 1905: 307).

The "biometricians," W. F. R Weldon and Karl Pearson, on the other hand, argued that new species could be identified with gradual changes over time in the normal distributions of various traits. Within a single population, for instance, natural selection could result in a dimorphic population that produce a "double-humped" curve, which then eventually transformed into two normal curves (Magnello 1996: 58). For the biometricians, speciation was conceived statistically, as were the species taxa themselves. What constituted a species was a distribution of variation that was statistically distinct – had a distinct normal curve – from the distribution of variation in other populations.

By the early middle of the twentieth century we begin to see a return to interest in species, in particular with the work of the architects of the "Modern Synthesis." Theodosius Dobzhansky devoted substantial attention to the species problem in his 1937 *Genetics and the Origin of Species*. There he listed the standard criteria in use at the time: morphological distinctness or lack of intergrades, geographic range, and sterility of hybrids (Dobzhansky 1937: 309–310). But these criteria, according to Dobzhanksy, seem to fail when taken singly. Some good species apparently have fertile hybrid offspring, while others have sterile hybrid offspring, but exhibit little morphological difference or distinction. Nor did it seem that a combination of criteria would work:

> Writers inclined toward eclecticism prefer to believe that none of the above criteria are sufficient when taken singly, but that a satisfactory result may be obtained by combining them. Of late, the futility of attempts to find a universally valid criterion for distinguishing species has come to be fairly generally, if reluctantly, recognized. This diffidence has prompted an affable systematist to propose something like the following definition of a species: "a species is what a competent systematist considers to be a species." (Dobzhansky 1937: 310)

But then, echoing Darwin, Dobzhansky argued that we should conceive of species as part of the historical, evolutionary process that produces evolutionary divergence. Part of this process, according to Dobzhansky, includes the development of *physiological* reproductive isolation. Quoting himself from an earlier work, he explained:

> In reality, discrete groups of organisms frequently coexist in the same territory without losing their discreteness, because their interbreeding is prevented through one, or a combination of several physiological isolating mechanisms ... The development of the latter causes a more or less permanent fixation of the organic discontinuity ... The stage of the evolutionary process at which this fixation takes place is fundamentally important, and the attainment of this stage by a group of organisms signifies the advent of species distinction. The present writer has therefore proposed ... to define species as that stage of the evolutionary process, "at which the once actually or potentially interbreeding array of forms becomes

segregated in two or more arrays which are physiologically incapable of interbreeding." (Dobzhansky 1937: 312)

Following this passage, he emphasized the dynamic nature of species (emphasis added):

> The definition of species just quoted differs from those hitherto proposed in that it lays emphasis on the dynamic nature of the species concept. *Species is a stage in a process, not a static unit.* This difference is important, for it frees the definition of the logical difficulties inherent in any static one. (Dobzhansky 1937: 312)

This way of thinking about species is significant in several ways. First, this is a historical or diachronic way of thinking about species. It is within the diachronic processes of speciation and divergence that Dobzhansky is asking us to conceive species – not primarily as static groups of organisms at some particular time. Second, and equally important, he followed Darwin in focusing on the *processes* that produce divergence. Reproductive isolation, for instance, was crucial to this divergence. Third, these processes were typically not directly observable but indirectly inferred:

> Although in separating species the systematists, with rare exceptions, have no direct information on the ability of the forms concerned to interbreed, the criteria used by them are capable of producing indirect evidence bearing on this point. (Dobzhansky 1937: 313)

The final thing to notice is that Dobzhansky interpreted this as implying that asexually reproducing organisms do not come in species groupings (Dobzhansky 1937: 319). The species category applies only to sexually reproducing organisms!

Other architects of the synthesis followed Dobzhansky's lead on many of these issues. Julian Huxley, for instance, in the introduction to his 1940 *New Systematics*, agreed with Dobzhansky on the importance of a dynamic conception of species as "stages in a process of evolutionary diversification" (Huxley 1940: 17). And while he focused on reproductive isolation, he wanted to place somewhat more emphasis on other factors than Dobzhansky:

> [T]here is no single criterion of species, Morphological difference; failure to interbreed; infertility of offspring; ecological, geographical or genetical distinctness – all those must be taken into account, but none of them singly is decisive. Failure to interbreed or to produce fertile offspring is the nearest approach to a decisive criterion. (Huxley 1940: 11)

Ernst Mayr also followed Dobzhansky's lead in asking us to think about species diachronically and dynamically, as a historical changing entity.

> If there is evolution in the true sense of the word, as against catastrophism or creation, we should find all kinds of species – incipient species, mature species, and incipient genera, as well as all the intermediate conditions. To define the middle stage of this series perfectly, so that every taxonomic unit can be certified with confidence as to whether or not it is a species, is just as impossible as to define the middle stage in the life of a man, mature man, so well that every single human male can be identified as boy, mature man, or old man. It is therefore obvious that every species definition can be only an approach and should be considered with some tolerance. (Mayr 1942: 114)

As we shall see later, there is a highly significant suggestion in the analogy between species and individuals, but the more obvious implication is also instructive. We can look at species synchronically, at a particular time, but this is only a small part of what it means to be a species, and may not be representative of the particular species as a whole. When observing organisms in nature at a particular time, we are observing only a single stage in the dynamic process of evolution. We cannot *define* a species taxon by looking at only a single stage, any more than we can *define* a person by looking at a single stage in his or her life.

Mayr also followed Dobzhansky, perhaps more closely than Huxley did, in focusing on reproduction in formulating a definition of the species category. After outlining five concepts he claimed were in use by taxonomists – practical, morphological, genetic, sterility and biological concepts – he asserted the priority of the "biological" concept based on interbreeding, because of its theoretical superiority. (Mayr 1942: 115–119). But then he admitted that for practical matters other concepts are required:

> A biological species definition, based on the criterion of crossability or reproductive isolation, has theoretically fewer flaws than any other. In practice, however, it breaks down just as quickly. Like them, it is not applicable to the isolated forms and these are the really important ones. As long as populations are in contact with one another, it is generally not difficult to arrive at a decision as to whether or not they are conspecific; the isolated forms are the ones that puzzle us. (Mayr 1942: 120)

Even though crossability or interbreeding cannot be determined in "alloptric" populations that do not overlap, Mayr still defined species in these terms:

A species consists of a group of populations which replace each other geographically or ecologically and of which the neighboring ones intergrade or interbreed wherever they are in contact or which are potentially capable of doing so (with one or more of the populations) in those cases where contact is prevented by geographic or ecological barriers ... Or shorter: species are groups of actually or potentially interbreeding natural populations, which are reproductively isolated from other such groups. (Mayr 1942: 120)

He repeated this definition in his 1982 *The Growth of Biological Thought* in three versions. The first version is in terms of actual or potential interbreeding:

Species are groups of actually or potentially interbreeding natural populations which are reproductively isolated from other such groups. (Mayr 1982: 273)

The second is in terms of reproductive communities:

A species is a reproductive community of populations (reproductively isolated from others) that occupies a specific niche. (Mayr 1982: 273)

The third version is in terms of the reproductive isolating mechanisms, defined as follows:

Isolating mechanisms are biological properties of individuals which prevent the interbreeding of populations that are actually or potentially sympatric. (Mayr 1982: 274)

Whether or not these are all equivalent formulations of an interbreeding criterion, Mayr does not think we can get by with just an interbreeding criterion. Other criteria are important for what they reveal about interbreeding (emphasis added):

The conspecifity of allopatric and allochronic forms, which depends on their potential capacity for interbreeding, *can be decided only by inference*, based on a careful analysis of the morphological differences of the compared forms. This does not mean that I am retracing my steps and now propose to accept a morphological species definition; no, it means simply that we may have to apply the degree of morphological difference as a yardstick in all those cases in which we cannot determine the presence of reproductive isolation ... To the supporter of a biological species concept, the degree of morphological difference is simply considered as a clue to the biological distinctness and is always subordinated in importance to biological factors. (Mayr 1942: 121)

What is most important here is not whether Mayr has equivocated in his formulations of a species concept, but that not all species concepts function in precisely the same way. Crossability or the tendency to interbreed is theoretically most important, and other criteria based on morphology, for instance, are significant for what they reveal about the tendency to interbreed.

George Gaylord Simpson followed Dobzhansky, Huxley and Mayr in many respects, but he also diverged from them in an important way. He agreed with them in recognizing a diachronic, historical conception of species in a 1951 paper in the Journal *Evolution*, and then again in his 1961 book *Principles of Animal Taxonomy*. In the book, he contrasted the non-historical interbreeding criterion of Mayr, or as he called it the "genetical concept," with a more historical "evolutionary concept."

The modern biological, or more strictly, genetical concept of a species among biparental, contemporaneous organisms is a group of interbreeding organisms genetically isolated from other such groups. That is a special case of a more extensive evolutionary concept of the species as a lineage with separate and unitary evolutionary role. The genetical species closely approximates a temporal cross section of such a lineage of biparental populations. (Simpson 1961: 147)

On this *evolutionary species concept*, species are lineages with distinct evolutionary fates, roles and tendencies: "An evolutionary species is a lineage (an ancestral-descendant sequence of populations) evolving separately from others and with its own unitary evolutionary role and tendencies" (Simpson 1961: 153).

So far this is in line with the views of Dobzhansky, Huxley and Mayr. But notice that Simpson did not refer *at all* in his species definition to the standard criteria of morphology, interbreeding, fertility/ sterility, geographic location – as his predecessors did. The conditions that made a group of organisms a species were instead formulated in terms of a lineage *"evolving separately from others and with its own unitary evolutionary role and tendencies."* These are clearly not the observable, *practical* criteria long used by naturalists from Cesalpino and Ray, to Linnaeus, Darwin and Dobzhansky and Mayr. Instead, they are what we might term *theoretical* criteria – criteria that are to be understood in terms of a theory – in this case evolutionary theory. The first condition, to be "evolving separately from others," seems to refer to the divergent change association with the branching of evolution in the above diagram. The second condition, to "have a unitary role and tendencies," is less clear. In part it is to be understood in terms of fit within an environment: "Roles are definable by their equivalence to niches, using 'niche' for the whole way of life or relationship to the environment of a population of animals" (Simpson 1961: 154). To have a unitary role here might involve having a unique niche or way of life relative to an environment.

What is important in Simpson's definition is that the traditional species criteria are useful, but only insofar as they have *theoretical significance* relative to divergent change and "unitary roles." Interbreeding, for instance, will be significant *just insofar* as it is theoretically significant – relative to divergence and the maintenance of a unique evolutionary role:

> The amount of interbreeding allowed by definition is then precisely as much as as does not cause their roles to merge. The taxonomic value of the genetic criteria of interbreeding and isolation lies not in those characteristics in themselves but in their evidence as to whether populations are or are not capable of sustaining separate and unitary roles over considerable periods of time. Interbreeding helps to keep a role unified; isolation makes possible separation of roles. (Simpson 1961: 153)

Similarly, morphology is significant, but *just insofar* as it has the requisite theoretical significance.

> Morphological resemblances and differences (as reflected in populations, not individuals) are related to roles if they are adaptive in nature. The assumptions that overall resemblance and difference is, on balance or as an average, adaptive is adequately justified by general evolutionary theory ... The definition requires only that the roles be separate but each unified, and that is as a safe enough rule, shown by somatic differences and resemblances between populations (Simpson 1961: 154).

This subservience of the traditional practical criteria to the theoretical is a highly significant move, and we shall return to it in the next chapter. But for now, the important thing to note is that we still have multiple species criteria that can be applied in different ways by different researchers. This is reflected in the work of the biologists and systematists that followed these founders of the Modern Synthesis.

AFTER THE SYNTHESIS

A casual observer could be excused for thinking that these architects of the synthesis made real progress in solving the species problem. After all, Mayr's *biological species concept* is the concept most often taught in introductory biology courses. If asked about the nature of species, many or even most non-systematists might give a definition in terms of one of Mayr's three versions – in terms of actual or potential interbreeding, reproductive community or reproductive isolation mechanisms. But for systematists the recent history has been cause for considerably less optimism. Because relatively few systematists are willing to follow Dobzhansky in concluding that the species category does not apply to asexual organisms, and because few see Mayr's *biological species concept* as fully adequate – given that it fails to apply to asexual organisms – there is little optimism that the solution is to be found in this concept. Furthermore, whereas Mayr identified five main species concepts in 1942 – practical, morphological, genetic, sterility and biological – more recent counts are typically in the twenties. Richard Mayden, for instance, has identified and distinguished at least twenty-two species concepts in circulation. If Mayden is correct about the count, or even close, then it seems that, instead of a resolution to the conflict between species concepts, we have a proliferation of conflicting concepts. The species problem is getting worse! There are currently, in very general terms, three kinds of species concepts, the first based on biological or evolutionary processes; the second based on various kinds of similarity; and the third based on the diachronic, historical component of species.

The basic idea behind the "process" species concepts is that we can identify and individuate species taxa on the basis of those biological or evolutionary processes that are responsible for the development or maintenance of species. Since Mayr's basic statement of the biological species concept in 1942, interbreeding and reproductive isolation have figured prominently in modern process concepts. It should be remembered, though, that he was endorsing criteria that had been in use for at least a couple of centuries. Recall the words of Buffon from chapter 3:

> We should regard two animals as belonging to the same species, if, by means of copulation, they can perpetuate themselves and preserve the likeness of species; and we should regard them as belonging to different species if they are incapable of producing progeny by the same means. (Lovejoy 1968: 93)

Modern systematists have seen reproduction as responsible for the cohesion within a species, and the maintenance of the barriers between species.

In the first of Mayr's three versions of his *biological species concept*, he appealed to actual or potential interbreeding and reproductive isolation: "Species are groups of actually or potentially interbreeding natural populations which are reproductively isolated from other such groups" (Mayr 1982: 273). This way of thinking about reproduction is in terms of populations – and in the reproductive isolation that develops between populations. This has evolutionary significance in that divergent change seems to require the elimination of reproductive blending. But in the third formulation of the *biological species concept*, Mayr focused on the properties of individuals that prevent interbreeding. "Isolating mechanisms are biological properties of individuals which prevent the interbreeding of populations that are actually or potentially sympatric" (Mayr 1982: 274). On this approach to reproduction, it is not so much the reproductive behavior and success in populations that is used as a criterion, but the physiological basis for that behavior.

Hugh Paterson has followed up on that idea in his *recognition species concept*, that species are populations of individuals which share a common specific mate-recognition system (Paterson 1993: 105; Mayden 1997: 408). According to this species concept, the organisms of a single species have cohesion due to a *specific mate recognition system* (SMRS) that consists of a series of signaling methods, including, among other things, chemicals which determine mating patterns, as well as the physiological compatibility of the sexes required to mate. This is important because sometimes the question is not whether two organisms could *potentially* produce offspring, but whether they recognize each other as potential mates and behave accordingly. Paterson claims that this is a "cohesion" concept and contrasts it to Mayr's "isolation" concept, emphasizing the reproductive cohesion we find in species, rather than the isolation – the discontinuity – from other species. While some see this as only a difference in emphasis, others see it as a highly significant advance over the *biological species concept*. Niles Eldredge, for instance, in a recent volume devoted solely to the *recognition concept*:

> It is my conviction that Hugh Paterson's concept of the *Specific-Mate Recognition System* ... has contributed more to the difficult task of sorting competing claims on the nature of species – and related issues – than

any other formulation that has appeared since the origins of the "modern synthesis" in the late 1930s and early 1940s. (Eldredge 1995: 464)

Eldredge sees this concept as significant first, in that it gives a criterion that can be applied to allopatric populations – populations of organisms that don't occupy the same place, and second, in its significance for speciation. The idea lurking behind the *recognition species concept* is that new species form on the basis of modification of the specific mate recognition systems, and isolation is therefore only a byproduct of the divergence of mate recognition systems (Eldredge 1995: 467).

Paterson also sees his concept, based on *specific mate recognition systems* (SMRS), as relevant to another way of thinking about species – in terms of a gene pool: "It is evident that a new SMRS ... determines a new gene pool and, hence, a new species" (Paterson 1993: 33; see also Mayden 1997: 409). Since the beginning of the synthesis the emphasis has sometimes been less on reproductive behavior than on the underlying population genetics. Dobzhansky had asked us to think about biological taxa in precisely these genetic terms. The *genetic species concept* follows up on this suggestion and is based on the idea that there is a reproductive cohesion in a population due to the common genetic foundation that makes reproduction possible. Mayden attributes this concept to Dobzhansky whom he quotes about species as "the largest and most inclusive reproductive community of sexual and cross-fertilizing individuals which share in a common gene pool" (Mayden 1997: 399). This idea is based, as is the recognition concept, on the idea of a reproductive community. But rather than looking at whole organisms and their mate recognition systems, it turns to the genetic variation found in the organisms that make up such a community.

The reproductive concepts just outlined are limited in applicability and scope – a species concept based on interbreeding can apply only to biparental, sexual species. The *agamospecies concept* was proposed to serve as an umbrella concept for all taxa that are uniparental and asexual, as Mayden explains:

> This concept refers specifically to taxa that do not fit the biparental, sexually reproducing mode. It serves as a general umbrella concept for all taxa that are uniparental and reproduce via asexual reproduction; often these species are the result of interspecific or intergeneric hybridization. These species may produce gametes but there is often no fertilization, except via hybridization. (Mayden 1997: 389)

This concept allows us to disagree with Dobzhansky, who argued that asexual organisms did not form species since they lacked the cohesion of reproduction, and agree with Simpson who argued that there were other sources of cohesion based on the ability for genes to spread within a population on the basis of natural selection (Simpson 1961: 162–163).

While reproduction has been perhaps the most prominent biological process incorporated in species concepts, other processes based on geographic and ecological factors have been invoked as well – both implicitly and explicitly. Geographic discontinuity, as we saw at end of the last chapter was the basis for species determinations in Darwin's time. It was also plausibly implicit in Mayr's second version of the *biological species concept*: "A species is a reproductive community of populations (reproductively isolated from others) that occupies a specific niche" (Mayr 1982: 273). Although Mayr thought that *biological isolation* (based on the biological properties of organisms such as physiology and behavior) was more important than *geographical isolation* (Mayr 1982: 274), he thought that biological isolation often began and developed as a product of geographic isolation (Mayr 1942: 180). And although Dobzhansky also thought that geographical isolation ultimately required physiological isolation, he claimed that geographic isolation was an important criterion for many (Dobzhansky 1937: 230). Mayden does not explicitly recognize such a *geographic species concept*, but we see such an idea in the views of Dobzhansky and Mayr. It was also a criterion recognized by Darwin and his contemporaries, as we saw at the beginning of this chapter.

Also found in Mayr's second version of the biological species concept is the idea of a *niche*. This idea might plausibly be seen as operating in the *ecological species concept*, proposed by Van Valen, which asserts that a "species is a lineage (or closely related lineages) which occupies an adaptive zone minimally different from that of any other lineages in its range" (Van Valen 1992: 70; see also Mayden 1997: 394–95). Here what counts as a species depends on the adaptive zone. An adaptive zone is, according to Van Valen:

> [S]ome part of the resource space together with whatever predation and parasitism occurs on the group considered. It is a part of the environment, as distinct from the way of life of a taxon that may occupy it, and exists independently of any inhabitants it may have. (Van Valen 1992: 70)

Species are, in some sense then, ecological units. The significance of this approach is that the divergent change of species is often understood to be related to an adaptive zone, and the adaptive zone is integrative in the sense that it maintains a common adaptive response among the individuals of the species.

These process concepts have been criticized for the difficulty in applying them to nature. After all, how can the potential for interbreeding be determined, especially with groups of organisms that are geographically isolated? What can be applied, to varying degrees, are those species concepts based on similarity. Among these similarity concepts are the *morphological species concept*, which has been in use from long before Darwin to the present, and asserts that "species are the smallest groups that are consistently and persistently distinct, and distinguishable by ordinary means" (Mayden 1997: 402). According to this concept, which is based on morphological distinctness – the gaps between groups of organisms, and degree of difference – we can group organisms on the basis of observed physical traits such as skeletal structures, sexual organs, size, shape, beak length etc. While this is perhaps the easiest criterion to apply – as Darwin noted – it does not come without problems. It is neither necessary nor sufficient. The first and most obvious problem is the appearance of "cryptic species," groups of organisms that have no obvious morphological discontinuity or difference with other organisms, yet nonetheless do not reproduce. In this case, morphological distinctness is not necessary. The second problem is that this concept is not sufficient. Variability within species may present gaps and difference that no one would identify as species determinative differences. The tiny male barnacles that Darwin thought were parasites on the much larger females, for instance, would be members of a different species on a morphological criterion. But this would be unsatisfactory even for those who swear by the *morphological species concept*. Nor would anyone accept the morphological differences in the various developmental stages of butterflies, for instance, as species criteria.

These problems don't seem to be solved by the *phenetic species concept*, that asserts that overall similarity is the correct grouping criterion, and "the species level is that at which distinct phenetic clusters can be observed" (Mayden 1997: 404). On this concept, organisms are grouped into species on the basis of multiple features using algorithms designed to establish overall similarity. This approach has the same problems as the morphological species concept, but it has additional problems in that different algorithms and choices of features produce different overall similarity groupings (Richards 2002). Similarly, the *polythetic species concept* is based on similarity, but is more straightforwardly a "cluster concept that defines species in terms of significant statistical covariance of characters" (Mayden 1997: 408). On this approach, some organisms may be grouped together into species on the basis of the possession of some subset of similarities, and different organisms may be grouped on the basis of some *other* subset.

We can also think of similarity in terms of molecules, in particular genes and DNA. On the *genotypic cluster concept*, species are "clusters of monotypic or polytypic biological entities, identified using morphology or genetics, forming groups of individuals that have few or no intermediates when in contact" (Mayden 1997: 400). Similarly, the *genealogical concordance concept* asserts that "population subdivisions concordantly identified by multiple independent genetic traits should constitute the population units worthy of recognition as phylogenetic taxa" (Mayden 1997: 397). The same problems that plague concepts based on morphological similarity, apply to concepts based on molecular similarity. First, different systematists group on the basis of different molecular characters and employ different similarity algorithms and clustering approaches. But more important, because some similarities are more significant than others, raw genetic similarity simply does not seem to cut nature at its joints. For instance, because some small genetic differences can have disproportionately large effects, we may not want to classify simply on overall genetic similarity. Not all genetic similarities and differences are equal in evolution and development. Furthermore, the concepts based on molecular similarity do not distinguish polymorphisms – variability – within reproductive communities, from differences among communities. In particular, they might well treat genetically based differences between the sexes the same as differences among species. Should we classify dimorphic, interbreeding males and females into different species? Finally, a raw similarity-based concept, based on either traditional morphology or molecular similarity, does not distinguish those similarities that are due to common ancestry (homologies) from those due to parallel evolution (homoplasies). This last worry about similarity approaches is highlighted in consideration of the third kind of species concepts based on the historical, diachronic dimension of species.

As we have seen in chapter 3 and the first part of this chapter, the historical dimension of species has been recognized in most species discussions both before and since Darwin. John Ray, Linnaeus and Buffon all conceived species as having a temporal dimension, and in terms of a lineage. Darwin was even more explicit, referring the reader to his branching tree that represents the evolution and branching of lineages. As we saw, this idea that species have a historical component, and change over that history, was also embraced by the architects of the synthesis: "Species is a stage in a process, not a static unit" (Dobzhansky 1937: 312).

This historical way of thinking about species is well represented in more recent species concepts. E. O. Wiley and Mayden adopted Simpson's suggestion and term – the *evolutionary species concept* (ESC). They quote Simpson, then give their own statement of the concept, arguing that a species is "an entity composed of organisms which maintains its identity from other such entities through time and over space, and which has its own independent evolutionary fate and historical tendencies" (Mayden 1997: 395; see also Wiley and Mayden 2000). Other historical concepts include the *successional species concept*, used for identifying fossil taxa. The idea here is that the change inferred in a lineage from fossil remains can be subdivided into multiple successive species segments. Similarly, the *paleospecies concept* and the *chronospecies concept* each conceive of species as segments of a changing lineage (Mayden 1997: 410–411). Because the process of drawing lines in the gradual transformation of a lineage has often seemed problematic, some systematists focus on speciation. The *cladistic species concept*, the *composite species concept*, the *internodal species concept*, and the *phylogenetic species concept* are all based on the idea that speciation events can serve to demarcate the beginnings and endings of species lineages. Mayden quotes Ridley on the *cladistic species concept*, which is associated with "cladistic" systematics: species are "that set of organisms between two speciation events, or between one speciation event and one extinction event, or that are descended from a speciation event" (Mayden 1997: 391). Similarly, the *composite* and *internodal* concepts are based on speciation – more specifically on the branching of lineages associated with speciation (Mayden 1997: 393, 401). The *phylogenetic species concept*, for which Mayden distinguishes three versions, follows the cladistic concept, in being based on the theory and operational method of cladistic (or "phylogenetic") systematics (Mayden 1997: 405–408; Mishler and Theriot 2000; Wheeler and Platnick 2000). In each of these versions, it is assumed that species taxa consist of individual organisms that are linked in a phylogenetic pattern of ancestry and descent.

THE SPECIES PROBLEM NOW

Since Mayden's accounting of species concepts in 1997, little has changed. Systematists still have not come to a consensus about the nature of species, and continue to use different species concepts with conflicting outcomes. A recent article, "What Is a Species?" by science journalist Carl

Zimmer in the popular science magazine *Scientific American*, highlights the ongoing species problem. Zimmer observes: "It may come as a surprise to see scientists struggling to agree on something so basic as how to decide that a group of organisms forms a species ... But in fact, the very concept of species has fueled debate for decades" (Zimmer 2008: 74). While Zimmer is clearly wrong about how long the debate has gone on – understating it by centuries, if not millennia, he is right in that the species problem has not been solved by our increasing understanding of biodiversity and evolution.

> What makes this disagreement all the more remarkable is that scientists now know vastly more about how life evolves into new forms than when the debate first started. Not long ago taxonomists could only judge a new species based on what they could see – things like fins, fur, and feathers. Today they can read DNA sequences, in which they are discovering a hidden wealth of biological diversity. (Zimmer 2008: 74)

Nonetheless, Zimmer suggests that a solution might be forthcoming. He describes a "unified approach": "Because of the turmoil, some researchers have begun to create phylogenetic classifications by looking beyond evolutionary history and combining it with molecular, ecological, behavioral and biological data" (Zimmer 2008: 78). Zimmer then goes on to describe a case where researchers used evolutionary history, ecological role and gene sequences to group a genus of spiders, *Promyrmekiaphila*, into six species. But it should be obvious by now that this will not solve the species problem – given that each criterion might divide biodiversity differently.

A unified approach that merely combines criteria, does not tell us, first, how to resolve conflicts between criteria; second, which of the twenty-plus criteria should be employed; and, third, which criteria should be more heavily weighted. We may be committed to taking evolutionary history into account, but that does not by itself tell us which historical species concept to use. Should we employ the *evolutionary species concept*, the *cladistic species concept* or one of the three *phylogenetic species concepts*? We might also use the *paleo* or *chronospecies* concepts. We cannot, however, use all of these historical concepts because they have different implications for species groupings. Similarly, we could use similarity-based concepts, but which one: a *morphological, polythetic, phenetic* or *genotypic cluster* concept? It is one thing to say that we should combine concepts into a unified approach, it is another to say how it can be done consistently and unequivocally.

111

This desire to work out a unified approach is important in that it reflects the value of unification in science. As philosophers of science have often noted, science works in part by the unification of phenomena under concepts and laws. How this might be so, and how it might be applied to the species problem, is the topic of the next chapter. As we shall see, there is a way to combine the different species concepts into a single unified approach. Such an approach, however, will require that we rethink the fundamental nature of our species concepts and, more importantly, the conceptual framework in which these concepts function.

5

The division of conceptual labor solution

The three preceding chapters have traced the history of the species problem, revealing both the long endurance of the species problem and the inadequacy of the Essentialism Story. There have been multiple ways of conceiving species, from the enmattered forms and logical universals of Aristotle to the morphological, genealogical, reproductive and geographic conceptions of modern systematists that divide up biodiversity in different ways. The thinking about species that preceded Darwin was clearly not dominated by the property essentialism of this story. But what is also striking in this history are the post-Darwinian advances in our empirical understanding of the living world. We know much more now about the processes governing speciation, development, heredity and biogeography than in Darwin's day. But instead of solving the species problem, it seems to have made it worse – in the proliferation of new species concepts. What we now know about the molecular basis of heredity, for instance, has prompted the formulation of additional species concepts based on various kinds of genetic similarity.

How should we respond to all this? We might conclude that the species problem is not ultimately an empirical problem; it is not due to a lack of knowledge about biodiversity. This will be one of the conclusions of this chapter. Instead the species problem is found in the *conceptual framework* in which species concepts function. And as we shall also see, the solution to the species problem is likewise found in the conceptual framework. But there are other ways we might respond to the species problem. We might be skeptics, concluding that we cannot arrive at a satisfactory conception of the species category, either because there are no species things at all, or there is no *single kind* of species things. One skeptical response is *nominalist*, denying that species names pick out *any* real things. Another skeptical response is *pluralist*, denying that species names all pick out one and the same kind of things. On both responses, the search for a single way to conceive species is misguided and bound to fail. While I will be arguing here for a positive solution to the species problem, given the lack of progress we have seen in the long history of the species problem, these skeptical responses should be taken seriously.

The most skeptical of responses to the species problem is commonly described as "nominalist," a term that gets applied to a variety of positions. As we saw in the previous chapter, Darwin and Buffon both seemed nominalist in some of their writing, although there were also good reasons to doubt that they were true nominalists. There was at least one true nominalist, described in chapter 2, in the eleventh century thinker, Roscelin of Champiègne, who argued that universal terms were mere vocal utterances. On this radical view, species and genus terms were spoken sounds applied arbitrarily to things in the world. The terms *man* and *animal* are just words that designate individual things sharing nothing more than a common name. The biological implausibility of this radical nominalism is obvious. Surely humans have common features that distinguishes them from non-humans. The fact that Socrates and Plato are both human can hardly be nothing more than a mere consequence of linguistic practice. It is hard to see how this skeptical response is open to anyone who accepts modern biological thinking.

More plausible is the nominalism better described as "conceptualism" and associated with Peter Abelard and William of Ockham. On this approach, universal terms are to be associated with concepts in the mind, but these concepts nonetheless reflect real features of the world. For Abelard, as we saw in chapter 2, concepts were products of abstraction, but caused by similarities in the world. So while the universal terms *man* and *animal* signified concepts, not things in the world, they were nonetheless grounded on the real similarities among humans and animals. And for Ockham, there were three levels of language – written, spoken and conceptual. Written and spoken languages were dependent on conceptual language, a mental language that *naturally* signifies real similarities in nature. For both of these thinkers, universals – *species* and *genera* – signified something mental – a concept, but nonetheless reflected real features of the world.

There is something to be said for the recognition by Abelard and Ockham of the active nature of concept formation and use. We will return to that topic later in this chapter. But the main shortcoming of conceptualism, relative to the species problem, is that it was not a *biological* approach. The medieval debate about universals was about the nature of predication in general, not about the division of the living world into species and genera on the basis of important biological features. What that means is that concept formation could be based on and reflect real features of the world, but nonetheless not pick out species and genera in the biological, taxonomic sense. The species and genera of the medieval conceptualists were general categories that we applied to all things in the world – not just biological. But even if we did focus more narrowly on the relevant biological features, we could still divide the world conceptually in different ways. This is the idea behind the second skeptical response based on *pluralism*: species terms apply to real things and features of the world, but not a single kind of thing.

SPECIES OF PLURALISM

One form of pluralism is pragmatic and in the spirit of the conceptualism of Abelard and Ockham. According to this approach, we group and divide living things into categories on the basis of practical interests. For instance, we group some plants into *houseplants*, and others into *weeds*. We group some animals into *pets*, others into *wild animals*, and yet others into *farm animals*. We have *green vegetables*, *red meats* and *tropical fruits*. In each of these cases, there are real, objective features that guide our grouping, based on, among other things, color and chemical functioning. But these groupings are also arbitrary in the sense that they reflect the pragmatic interests we have in making the grouping. If we did not have an interest in the nutritive function of green vegetables, there would be no good reason to distinguish green vegetables from non-green vegetables or from non-vegetables. If species terms were like the terms *green vegetables* and *red meats*, then we would surely be justified in a skeptical attitude towards species groupings. After all, there are other ways to group vegetables and meats that also reflect real interests. We might, for instance, treat french-fried potatoes as vegetables, but for purposes of ordering from a menu and not on the basis of nutritive functioning. In each case there is some feature of the world that governs a classification, so it is not entirely arbitrary, yet the relevant feature does not seem to legitimately ground *biological* species groupings and designations. Consequently, these pragmatically based concepts do not typically pick out things that we might regard as good biological species.

There might, however, be theoretical and biological interests that can legitimately ground species groupings, and potentially do that in different ways. This is the premise of *pragmatic pluralism*. On this approach, biodiversity should be conceived and divided up according to the interests guiding the classification. Philip Kitcher develops such an approach, dubbing it "pluralistic realism." According to Kitcher:

> Species are sets of organisms related to one another by complicated, biological interesting relations. There are many such relations which could be used to delimit species taxa. However, there is no unique relation which is privileged in that the species taxa it generates will answer to the needs of all biologists and will be applicable to all groups of organisms. (Kitcher 1992: 317)

There are, according to Kitcher, multiple ways to conceive species because there are multiple kinds of biological investigations, based on different theoretical interests. Kitcher identifies two main pragmatic subdivisions, structural and phylogenetic:

> The species category is heterogeneous because there are two main approaches to the demarcation of species taxa and within each of these approaches there are several legitimate variations. One approach is to group organisms by structural similarities. The taxa thus generated are useful in certain kinds of biological investigations and explanations ... The other approach is to group organisms by their phylogenetic relationships. Taxa resulting from this approach are appropriately used in answering different kinds of biological questions. (Kitcher 1992: 317)

We might adopt a structural approach, for instance, if we were interested in the genetic basis for the development of various traits or organs, or the common chromosomal structure or developmental programs (Kitcher 1992: 328–9). We might adopt the phylogenetic approach, on the other hand, if we were instead looking at evolutionary processes that produce continuities and divisions among organisms based on reproductive isolation, ecological and morphological distinctness (Kitcher 1992: 331).

The advantage of this *pragmatic pluralism* is obvious. It makes sense of the many different things biologists have focused on in their thinking about species, from morphology, genetics, DNA, interbreeding to ecological functioning:

> The existence of diverse species concepts is not altogether unexpected, because concepts are based on properties that are of the greatest interest

to subgroups of biologists. For example, biologists who study hybrid zones tend to emphasize reproductive barriers, whereas systematists tend to emphasize diagnosability and monophyly, and ecologists tend to emphasize niche differences. Paleontologists and museum taxonomists tend to emphasize morphological differences, and population geneticists and molecular systematists tend to emphasize genetic ones. (de Queiroz 2005: 6601)

The disadvantages of this pluralism should be obvious as well. Different *pragmatic* species concepts group into species taxa in different and inconsistent ways. The structural species of the developmental biologists are not the historical species of the phylogeneticist. What counts as a species then depends on the theoretical interests and needs of the investigators. And if there is no single set of theoretical interests, there can be no single grouping of species, and no single classification. One biologist with a particular theoretical interest can group one way, and another biologist with a different theoretical interest can just as legitimately group another way. But if we require a *single* species classification, the only obvious way to resolve the differences is political. We simply favor the theoretical interests of one group of biologists over the other. Surely this is problematic. In classification, we must then ask: whose species taxa get incorporated into the taxonomy? And in the application of endangered species law, we must ask: whose species taxa get preserved? The pragmatic pluralist, it seems, has to give some account of how to satisfy these sorts of demands without resorting to mere politics.

Another pluralist response, *ontological pluralism*, asserts that we have not been able to settle on a single species concept because the species category itself is heterogeneous and includes different kinds of things. In other words, there is no essence to the species category. This pluralism comes in a variety of forms. There is a relatively modest, *ranking pluralism*. According to one version of ranking pluralism based on a cladistic approach to classification, species are like higher-level groups (such as genera), in that they are "monophyletic," consisting of only an ancestor and all its descendents. But there is no single, universal criterion that distinguishes the species level from other levels in the branching of the evolutionary tree (Mishler and Brandon 1987). Mishler and Donoghue see this as a *ranking problem*:

Even when monophyletic groups are delimited, the problem of ranking remains since monophyletic groups can be found at many levels within a clade. Species ranking criteria could include group size, gap size, geological age, ecological or geographic criteria, degree of intersterility, tradition, and possibly others. The general problem of ranking is presently unresolved, and we suspect that an absolute and universally applicable criterion may never be found, and that, instead, answers will have to be developed on a group by group basis. (Mishler and Donoghue 1992: 131)

Mishler and Donoghue think that since the ranking of monophyletic groups into species is done on the basis of multiple criteria, we require multiple species concepts: "we think that a variety of species concepts are necessary to capture the complexity of variation patterns in nature" (Mishler and Donoghue 1992: 131).

The biggest problem with this proposal is that it is not obvious how to apply the concept of monophyly to the species level. It cannot be as usually formulated – *in terms of species*: a monophyletic group includes an ancestral species, all its offspring species, and no other species. Mishler and Donoghue recognize this is a problem:

Several different concepts of monophyly have been employed by systematists, but none explicitly at the species level ... We favor Hennig's concept of monophyly (except explicitly applied at the species level) but are fully aware of the difficulties in its application at low taxonomic levels ... In particular the difficulty posed by reticulation. (Mishler and Donoghue 1992: 130)

One question about species monophyly: does it require that, to count as a species, all *organisms* within the species must have a single common ancestral "founder" organism? The species would then be that founder organism and all its offspring. But then in sexually reproducing species there would presumably be at minimum two founding organisms. And the offspring of this pair would presumably be interbreeding with individuals that were not offspring of the pair. There would be reticulation. This proposal based on monophyly looks to be problematic if we must apply the idea of monophyly at the species level or lower.

Another ontological pluralism has been developed by Marc Ereshefsky. He accepts the basic idea that species are genealogical, historical lineages, but denies that they are all the same kinds of lineages. First he begins by noting there are three main ways of thinking about species – in terms of interbreeding, ecology and monophyly. Then he argues that these are all kinds of lineages produced by different evolutionary forces.

The positive argument for species pluralism is simply this: according to contemporary biology, each of the three approaches to species highlights a real set of divisions in the organic world ... All of the organisms on this planet belong to a single genealogical tree. The forces of evolution segment that tree into a number of different types of lineages, often causing the same organisms to belong to more than one type of lineage. The evolutionary forces at work here include interbreeding, selection, genetic homeostasis, common descent, and developmental canalization ... The resultant lineages include lineages that form interbreeding units, lineages that form ecological units, and lineages that form monophyletic taxa. (Ereshefsky 2001: 139)

These different kinds of lineage concepts apply in different ways to biodiversity. Some organisms, for instance, may not form ecological lineages. That lineage concept would therefore not apply (Ereshefsky 2001: 142). These concepts also cross-classify, in that a group of organisms may be part of two different kinds of lineages, and hence members of two different species – and two different *kinds* of species. There will also likely be a difference in species counts: "when surveying a plot of land we might find that the interbreeding approach identifies three distinct species whereas the phylogenetic approach identifies two speciea" (Ereshefsky 2001: 156). Even though these species lineages are all *lineages,* Ereshefsky does not think they can be reduced to a single, monistic way of conceiving lineages. He considers and then rejects the suggestion that there is a "fourth parameter" that can unite these three different kinds of lineages under one conception (Ereshefsky 2001: 145). To see if he is right about the impossibility of further unification, we need to look at yet another pluralistic approach.

HIERARCHICAL PLURALISM

The *hierarchical pluralism* advocated by Richard Mayden and Kevin de Queiroz postulates a hierarchy of different *kinds* of species concepts. Mayden argues that there are two kinds of species concepts: *primary theoretical concepts* tell us what kinds of things species taxa are; *secondary operational concepts* tell us how to identify and individuate species taxa. This approach is hierarchical because the operational concepts depend on the theoretical concepts. Operational concepts do not tell us what species are but, *given a particular theoretical concept*, how to identify and individuate them (Mayden 1997: 383). The criteria for evaluating

119

species concepts depend on the way they function. Primary theoretical concepts must be consistent with current science and be general, applying as widely as possible:

> What then are the criteria we should be looking for in a primary concept? It should be consistent with current theoretical and empirical knowledge of diversification. It should be consistent with the ontological status of those entities participating in descent and other natural properties ... Finally, it should be general enough to encapsulate all types of biological entities considered species as taxa by researchers working with supraspecific taxa. (Mayden 1997: 419)

In other words, a primary theoretical concept needs to be theoretically adequate and universal. Operational concepts, on the other hand, need to be applicable to biodiversity, yet still theoretically relevant, as determined by the primary theoretical concept. Operational concepts might do this through morphological similarity, genetic similarity, interbreeding, ecological functioning and so on.

Mayden argues that the one concept capable of serving this theoretical function is the *evolutionary species concept (ESC)*. Mayden gives three statements of this concept. The first, from G. G. Simpson, asserts that a species is "a lineage (an ancestral-descendent sequence of populations) evolving separately from others and with its own unitary evolutionary role and tendencies" (Mayden 1997: 395). The second statement, from Edward Wiley, identifies species as "a single lineage of ancestor-descendent populations which maintains its identity from other such lineages and which has its own evolutionary tendencies and historical fate" (Mayden 1997: 395). The third formulation, from Wiley and Mayden, is that a species is "an entity composed of organisms which maintains its identity from other such entities through time and over space, and which has its own independent evolutionary fate and historical tendencies" (Mayden 1997: 395). The advantage of the ESC, according to Mayden, is that it serves the theoretical function and is universal: "The ESC is the most theoretically significant of the species concepts; it accommodates all 'types' of species known to date and thus has the greatest applicability. As such the ESC can serve as a primary concept" (Mayden 1997, 396).

The disadvantage of the ESC is that it is not operational. One cannot just observe lineages of the relevant kind in nature. The ESC therefore requires other operational, species concepts:

> While the ESC is the most appropriate primary concept, it requires bridging concepts permitting us to recognize entities compatible with

its intentions. To implement fully the ESC we must supplement it with more operational, accessory notions of biological diversity – secondary concepts. Secondary concepts include most of the other species concepts. While these concepts are varied in their operational nature, they are demonstrably less applicable than the ESC because of their dictatorial restrictions on the types of diversity that can be recognized, or even evolve. (Mayden 1997: 419)

Secondary, operational concepts are those that can be readily applied to biodiversity, and are indicative of species lineages. Species concepts based on morphological or genetic similarity, for instance, can help identify lineages, since organisms within a single lineage will generally share morphological and genetic traits. Concepts based on processes such as reproductive isolation, mate recognition systems and ecological niches can also be used to identify lineages since these are processes that operate in the formation and persistence of lineages.

Much of recent thinking about species, according to Mayden, has confused these two ways of conceiving species.

It is not uncommon to find in discussions of species and species concepts researchers confusing empirical data used in the operation of recognizing a species with a conceptualization or definition of species. Empirical data can include such things as anatomy, morphology, genetics (DNA, proteins), behaviour, etc., all possibly evaluated and analysed in a variety of ways and a variety of methods. (Mayden 1997: 388)

The confusion of these two ways of thinking about species results in the discordance of biological taxa – *the species problem* (Mayden 1997: 389).

The hierarchical pluralism of Kevin de Queiroz agrees with Mayden's in that it recognizes two kinds of species concepts. There are the *species concepts* proper, which give the necessary properties of species and provide theoretical definitions. Then there are *species criteria*, which give contingent properties and are "standards for judging whether an entity qualifies as a species" (de Queiroz 1999: 60). Many of the species concepts in use are to be understood as species criteria rather than species concepts proper.

The species criteria adopted by contemporary biologists are diverse and exhibit complex relationships to one another (i.e. they are not necessarily mutually exclusive). Some of the better-known criteria are: potential interbreeding or its converse, intrinsic reproductive isolation ... common fertilization or specific mate recognition systems ... occupation of a unique niche or adaptive zone ... potential for phenotypic cohesion ... monophyly as evidenced by fixed apomporhies ... or the exclusivity of genic coalescence ... qualitative ... or quantitative ... Because the entities satisfying these various criteria do not exhibit exact correspondence, authors who adopt different species criteria also recognize different species taxa. (de Queiroz 1999: 60)

Each of these criteria are contingent ways to identify species, but do not specify the nature of species.

The correct theoretical concept of species, that gives the necessary properties to be a species, according to de Queiroz, is the *general lineage concept*.

[S]pecies are segments of population-level lineages. This definition describes a very general conceptualization of the species category in that it explains the basic nature of species without specifying either the causal processes responsible for their existence or the operational criteria used to recognize them in practice. It is this deliberate agnosticism with regard to causal processes and operational criteria that allows the concepts of species just described to encompass virtually all modern views on species, and for this reason, I have called it the general lineage concept of species. (de Queiroz 1999: 53)

In a later paper, de Queiroz describes this general theoretical concept as a "metapopulation lineage": "sets of connected subpopulations, maximally inclusive populations" (de Queiroz 2005: 6601). Notice here that de Queiroz is endorsing a view much like Ereshefsky's, except that he does not individuate different species concepts on the basis of the evolutionary processes that generate species lineages.

According to de Queiroz, this general lineage concept (or metapopulation lineage concept) is accepted by virtually all biologists, even though there is disagreement in the use of operational species criteria:

Although all modern biologists equate species with segments of population lineages, their interests are diverse. Consequently, they differ with regard to the properties of lineage segments that they consider most important, which is reflected in their preferences concerning species criteria. Not surprisingly, the properties that different biologists consider most important are related to their areas of study. Thus, ecologists tend to emphasize niches; systematists tend to emphasize distinguishability and phyly; and population geneticists tend to emphasize gene pools and the processes that affect them. Paleontologists tend to emphasize the temporal extent of species, whereas neontologists tend to emphasize the segments of species that exist in the present. (de Queiroz 1999: 65)

This disagreement about criteria, it should be noted, does not imply different theoretical species concepts, but different practical considerations. It is only when the operational criteria are mistaken for theoretical concepts that there will be disagreement, generating a species problem.

There is much more that can be said about these two versions of hierarchical pluralism, and as I shall be arguing, much to recommend them. But for now we can just note several important implications of this general approach. First, the fact that there are different *kinds* of species concepts that function in different ways, and that have different standards of evaluation, suggests that we need to rethink traditional approaches to the evaluation of species concepts. David Hull explores (without necessarily endorsing) this tradition. He identifies three main evaluative criteria for species concepts – theoretical significance, universality and applicability (Hull 1997: 357–380). Then he analyzes various concepts and concludes that, while some concepts did significantly better than others relative to a single criterion, no species concept seemed to be significantly better than all others when evaluated *relative to all three*. Hull consequently declines to recommend any single concept. What is significant is that this analysis assumes that a satisfactory species concept must be tested against *all these criteria*, and then be compared with other species concepts on this *same* set of criteria. On this strategy the evolutionary, biological, ecological, phenetic and other species concepts are regarded as equivalent, competing concepts to be evaluated on the same set of criteria. Given such an evaluative framework, it is not surprising that Hull came to a negative conclusion. Hierarchical pluralism suggests that we need to reject this evaluative framework, for one that recognizes the different ways species concepts function.

Second, once we understand the conceptual framework in which species concepts function, and how this functioning varies, we can see how to better evaluate species concepts and perhaps come to a satisfactory, unified account of the species concept that also serves the purposes of evolutionary theory and biological classification. A good solution would make sense of all the ways we have conceived, identified and individuated species, and preserve the evolutionary and taxonomic significance of the species category. The division of conceptual labor solution to the species problem, to be developed in this chapter, and implicit in these two versions of hierarchical pluralism can plausibly do so. To see how, we need to more systematically look at three philosophical commitments. The first is *consilience* or the conceptual unification of phenomena in science. The second is the division of conceptual labor itself – how different concepts can function in different ways in support of the same theoretical goals. The third is the evaluative role of the theoretical framework – the adequacy of theoretical concepts is dependent at least in part on how they function within the overarching conceptual framework.

WHEWELLIAN CONSILIENCE

The skeptical solutions to the species problem outlined above, from the radical nominalism to the ontological pluralism, all have a cost. Radical nominalism seems to make species groupings arbitrary and theoretically insignificant. Pragmatic pluralism seems to imply that claims about species taxa are subjective, based on individual interests of researchers. A group of organisms is a species taxon *only* relative to a particular theoretical interest. Ontological pluralism, while it doesn't imply there are no such things as species, or that species groupings are arbitrary, nonetheless seems to challenge the basis of biological taxonomy. In our classifications, we would be including different, and not necessarily equivalent, kinds of things in the species category. And if the species criteria cross-classifies, as Ereshefsky allows, then there are multiple, inconsistent groupings. It is not clear how on *any* of these skeptical solutions we can continue to claim that species are the fundamental units of evolution and taxonomy. A "monist" solution to the species problem that preserves the biological significance of the species category would surely be preferable. "Unification" of species concepts is a worthy goal.

It has long been commonplace for philosophers of science to see unification as an important and necessary process of science. Standard histories of science seem to support this view. We are told, for instance, that science had its origin in the ancient Greek world, in the unifying theories of Thales, Anaximander, Empedocles and others, who argued for fundamental explanatory principles of the cosmos based on its material constitution. Others, such as the Pythagoreans, appealed to formal unifying principles based on mathematics, geometry and rational forms. Sometimes this unification was "nomological," achieved through subsumption of phenomena by empirical or causal laws. The motions of the planets, for instance, were unified by the descriptive laws of Kepler. Newton further unified these planetary motions with the descent of stones on earth and the tides through his universal, inverse-square law of gravitation that postulated a single, centripetal gravitational force. Maxwell's unification of electrical and magnetic forces, Einstein's unification of space and time, and the current search for a "grand unified theory" of physics are other examples, among many, of this unifying tendency. Perhaps more relevant here is Darwin's unification of biological phenomena, from the patterns of similarity among organisms and the apparent adaptiveness of organs to environments, to embryological development and vestigial organs, by his theory of evolution. And in modern science we see unification in the laws of biochemistry, chemistry, electromagnetism and particle physics.

There has also been a long tradition of unification that focuses on *concepts* rather than laws. There was a conceptual unification in Plato's theory of forms, by the correspondence of changeable things in the natural world to unchangeable transcendent "forms." For Plato, individual horses were horses by virtue of a relation to an eternal idea or form of a horse understood through reason. In chapter 2, we saw it in Aristotle's use of logical universals to understand the diversity of the world, as experienced in perception. The *eidos* as a logical universal was required for knowledge of the many different *eide* as enmattered forms. In those who followed Aristotle, and worried about the status of universals, the unification of individual things occurred through the application of universal names. Socrates was an individual thing, but was united with other similar things by the application of the term *man*. This idea of conceptual unification was perhaps most carefully developed by the nineteenth-century polymath William Whewell, who described it as a "consilience of inductions."

Whewell began with what he called the "fundamental antithesis." This antithesis plays out in several ways – through the antithesis of thoughts and things, theories and facts, and ideas and sensations. In each of these antitheses there is a distinction between the subject that has knowledge of the world and the world as it is experienced. At the most fundamental level are thoughts and things:

> Our thoughts are something which belongs to ourselves; something which takes place within us; they are what *we* think; they are actions of our minds. Things, on the contrary, are something different from ourselves and independent of us; something which is without us; they *are*; we see them, touch them, and thus know that they exist; but we do not make them by seeing or touching them, as we make our thoughts by thinking them. (Whewell 1984: 139)

Knowledge, for Whewell, required a combination of thoughts and things:

> In all cases, Knowledge implies a combination of Thoughts and Things
> ... Without Thoughts there would be no connexion; without Things, there
> could be no reality. Thoughts and Things are so intimately combined in
> our Knowledge, that we do not look upon them as distinct. One single act
> of the mind involves them both. (Whewell 1984: 140)

The antithesis of theory and facts is based on the more fundamental antithesis of thoughts and things:

> The Antithesis of Theory and Fact implies the fundamental antithesis
> of Thoughts and Things; for a Theory (that is, a true Theory) may be
> described as a Thought which is contemplated distinct from Things and
> seen to agree with them; while a Fact is a combination of our Thoughts
> with Things in so complete agreement that we do not regard them as sep-
> arate. (Whewell 1984: 146)

Facts are a product of thought, but in conformity with things. Theories go beyond the facts and are "moulded" by thought.

Finally there is the antithesis of ideas and sensations. Ideas are pro-
vided by thought, and applied to the things experienced in sensations. At the most basic level, space, time and number are all ideas applied to sensation.

> We see and hear and touch external things, and thus perceive them by our
> senses; but in perceiving them, we connect the impressions of sense accord-
> ing to relations of space, time, number, likeness, cause, &c. Now some at
> least of these kinds of connexion, as space, time, number, may be contem-
> plated distinct from the things to which they are applied; and so contem-
> plated, I term them *Ideas*. And the other element, the impressions upon our
> senses which they connect, are called *Sensations*. (Whewell 1984: 147)

Whewell saw this antithesis as applicable throughout science, from the orbits of planets, to gravitation force and to the identification and distinction of different kinds of organisms:

> [W]e see two trees of different kinds; but we cannot know that they are
> so, except by applying to them our Idea of the resemblance and difference
> which makes kinds. And thus Ideas, as well as Sensations, necessarily enter
> into all our knowledge of objects: and these two words express, perhaps
> more exactly than any of the pairs before mentioned, that Fundamental
> Antithesis, in the union of which, as I have said, all knowledge consists.
> (Whewell 1984: 149)

These ideas, it should be noticed, are not the *objects* of thought (of John Locke and the early modern philosophers), but *laws* of thought, governing the ways we think about things. Our ideas give us guidance on how to organize phenomena. This is the basis for Whewell's views on "induction."

In Whewell's time, and at least since Newton, science was assumed to be inductive. But precisely what that meant was the subject of many disputes. In a famous debate with John Stuart Mill, Whewell laid out his ideas of what constitutes induction. Mill had argued that induction was an *inference,* the generalization from particular facts. It was the conclusion that all heavenly bodies traced out an elliptical orbit from the observation that one or more planets had elliptical orbits. But according to Whewell, induction was the *superinducement* of the general idea of an ellipse on the observations (Whewell 1984: 338). It was the application of the *idea* of the ellipse to planetary motion. Similarly, it was the application of the *idea* of gravity as an inverse-square centripetal force to planetary motion and the descent of stones on earth. The idea of gravity is a *rule* of thought because it tells us how to approach the phenomena – how to think about them and how to measure them. In modern terms, it gives us operational guidance. One important virtue of the idea of gravity is that it unifies the phenomena of the planetary orbits, falling objects on earth, and the tides. It does this through what Whewell calls a "consilience of inductions." There are two main processes at work in this consilience: the "colligation of facts" and the "explication of conceptions."

The colligation of facts is the application of an idea or concept to observed phenomena – to empirical facts. These facts may be simple, observational facts such as the phases of the moon or the places of the sun's rising and setting. Or they may be facts inferred from more directly observable facts such as those "connexions" expressed by Kepler's laws of planetary motion (Whewell 1984; 205–206). One way this colligation is accomplished is through prediction, broadly interpreted. The idea of gravity colligates planetary motion, because it can predict the location of each planet at particular times, and it can predict Kepler's laws of planetary motion. Similarly, the idea of gravity colligates the descent of objects on earth, because it can be used to predict the rate of descent (combined with the laws of aerodynamics).

The ideas or concepts that colligate the facts are developed through a process of clarification and development, or "explication." That the gravitational force is proportional to mass and not extension for instance, is a clarification. The determination of the gravitational constant by Henry Cavendish constitutes a development of the idea. This explication happens through the practice of science, as Whewell explained:

> The Explication of Conceptions, as requisite for the progress of science, has been effected by means of discussions and controversies among scientists; often by debates concerning definitions; these controversies have frequently led to the establishment of a Definition ... the essential requisite for the advance of science is the clearness of the Conception. (Whewell 1984: 254)

The conceptions must be more than just clear, they must also be appropriate for the subject matter. Whewell saw this as a requirement that the conception be an appropriate modification of the fundamental idea. He thought that, in mechanics, the fundamental idea is of *force*; in chemistry, it is of *substance;* in biology, it is of *life* (Whewell 1984: 197–199). In each of these domains, the fundamental ideas govern speculation and the invention of new conceptions. Scientific progress consists in the clarification and development of these fundamental ideas:

> [T]he Fundamental Ideas on which science depended, and the Conceptions derived from these, were at first vague, obscure, and confused; – that by gradual steps, by a constant union of thought and observation, these conceptions became more and more clear, more and more definite; – and that when they approached complete distinctness and precision, there were made great positive discoveries into which these conceptions entered, and thus the new precision of thought was fixed and perpetuated in some conspicuous and lasting truths. (Whewell 1984: 198–199)

For Whewell then, each domain of science had its fundamental ideas from which concepts are ultimately derived. This, in effect, required that a concept fit into the larger theoretical framework. The concept of gravity must fit into the framework that postulates forces. Explication is therefore driven not just by the empirical facts about phenomena, but by the demands of an overarching theoretical framework, and the fundamental ideas of the relevant domain.

The consilience of inductions fits into this framework of colligation and explication. A conception was adequate, for Whewell, if it was clear and colligates the facts it was devised to explain. But a conception was even better – "true" in Whewell's view – if as it was explicated, it came to colligate other kinds of facts.

> The *Consilience of Inductions* takes place when an Induction, obtained
> from one class of facts, coincides with an Induction, obtained from another
> different class. This Consilience is a test of the truth of the Theory in
> which it occurs. (Whewell 1984: 257)

While Whewell generally used examples from the most highly developed sciences of his day, physics and astronomy, this idea of consilience is
usually seen to apply also to the arguments of his contemporary, Charles
Darwin. In the last chapters of his *Origin*, Darwin argued that his theory
of evolution by natural selection should be regarded as true because it
makes sense of a variety of phenomena, from the patterns of similarity among organisms and classification, to embryological development,
biogeography, vestigial traits and co-evolution. Through the explication
of the concepts of natural selection and common descent, facts from all
sorts of different domain can be colligated. The fact that the theory of
evolution by natural selection unifies all this phenomena was taken by
Darwin to be a strong reason to accept it as true (Darwin 1964: 458;
Ruse 1998: 3).

My concern here is not whether Whewell got all the details of this
conceptual unification right. The main ideas can be profitably applied to
the species problem, even if the details were to be problematic. The first
useful idea is implicit in Whewell's fundamental antithesis of thought
and things, theories and facts and ideas and sensations. The point here
is that knowledge of the things in the natural world requires thought,
theories and ideas. We confront the world actively, with a set of concepts
that we develop and apply. Nature does not directly and simply provide
us with the concepts we need to understand it. This is Aristotle's insight
in his distinction between *eidos* as logical universal – a concept – and
eidos as enmattered form – a natural phenomenon. It was the insight of
the scholastics such as Abelard and Ockham who tried to understand
how we could get universal concepts, represented by universal terms,
from the complexity and variability of individual things we experience.
The first idea then is this: the scientific process depends, at least in part,
on the active and conscious invention, development and application of
concepts to phenomena.

The second useful idea, to be examined more closely in the next section, is that scientific concepts are "two-faced." They must serve two
masters – empirical facts and theoretical commitments. They are tested
by how they conform to facts about the world – how they *colligate* these
facts. Whether a scientific concept is adequate or not depends in part on

its success in its application to empirical phenomena. Some concepts can be more adequate as they apply to, and explain more phenomena. Then in the application of concepts to phenomena, the concepts get clarified and developed so as to better colligate the facts. It does this by unifying the facts under a single conception. But this explication of concepts is not just relative to a set of facts, but to an overarching theoretical framework that gives guidance as to how a concept can be developed. The concept of gravity, for instance, must be explicated preserving the assumption that it is a force. Similarly, the concept of a species must be explicated in a way that preserves its significance within evolutionary theory.

CONSILIENCE OF SPECIES CONCEPTS

The idea of consilience is problematic, however, when applied directly to the species problem – if the various species concepts are treated as equivalent. Michael Ruse has pursued this line of thinking, arguing that there is a developing consilience in species concepts: "There are different ways of breaking organisms into groups, and they *coincide!* The genetic species is the morphological species is the reproductively isolated species is the group with common ancestors" (Ruse 1992: 356). The problem with Ruse's proposal, as we have seen in previous chapters, is that it does not look as if this consilience is really forthcoming in a direct and simple manner. The genetic species is not always the morphological or reproductively isolated species. If there really were a developing consilience, then we would presumably *not* see the proliferation of species concepts that group organisms inconsistently. The treatment of the various species concepts as equivalent, functioning in the same way relative to the same empirical phenomena, seems to rule out this application of consilience. But if we apply the consilience idea to the hierarchical models of Mayden and de Queiroz, the prospects are more promising. Ruse's analysis may be on the right track, *if* we take into account the division of conceptual labor.

As we saw in a previous section, the hierarchical pluralisms of Mayden and de Queiroz are based on the idea that there are different kinds of species concepts that function in different ways. Some concepts are theoretical, telling us what species things are. Other concepts are operational, telling us how to identify species things. There are two ways we can potentially apply Whewellian consilience to this hierarchy. First we could apply it to both the operational and theoretical concepts. If so, the biological species concept can be seen to colligate some facts – facts about some sexually reproduction species. The agamospecies concept colligates other facts – facts about asexual species. The morphological species concepts colligate facts about some kinds of similarity. Eventually these concepts all get colligated by the superinducement of another, more inclusive concept – perhaps the *general lineage concept* of de Queiroz, or *the evolutionary species concept* of Mayden that conceives species in terms of population lineages.

One problem with this way of applying consilience is that the various species concepts seem to be based on different kinds of things and processes – morphology, reproductive cohesion and isolation, and so on. It is not obvious how a reproductive concept can be "explicated" to produce a historical or morphological concept. This would be conceptual replacement rather than explication. There is, however, another more promising way to apply the consilience idea to the species problem. On this alternate approach, consilience is appropriate for primary theoretical concepts, but not the secondary, operational concepts. To understand this application of consilience we need to look at the role of the relevant theoretical framework – the evolutionary framework – and how that informs the explication of theoretical species concepts.

THEORETICAL SPECIES CONCEPTS

At the most basic level, the theory of evolution tells us that there is change over time. In Darwin's theory of evolution this involved the origin of new species through divergent change. Darwin's principle of divergence, whereby varieties become species, is crucial (Darwin 1964: 111). It explained the branching evolutionary tree diagram that in turn served to illustrate Darwin's approach to classification.

> I request the reader turn to the diagram illustrating the action, as formerly explained, of these several principles; and he will see that the inevitable result is that the modified descendants proceeding from one progenitor become broken up into groups subordinate to groups ... So that we here have many species descended from a single progenitor grouped into genera; and the genera are included in, or subordinate to, sub-families, families and orders, all united into one class. (Darwin 1964: 412–413)

What is important here is that this tree emphasized the temporal, historical dimension of evolution, and the branching associated with speciation. It tells us that species have beginnings in speciation events. They have duration. They change. And they have endings. Since Darwin, this historical component has become further entrenched in evolutionary theory, in particular in the debates about the relation between microevolution – changes within a species, and macroevolution – changes across speciation events.

This is not to say, of course, that species taxa are *just* historical entities. Evolutionary theory tells us that they exist as well, as groups of organisms at particular times: groups that share similarities, sometimes interbreed, occupy ecological niches, vary geographically, form gene pools and have a variety of social structures. This way of thinking about species has been developed and refined most notably by the thinkers of the Modern Synthesis, such as Mayr, Dobzhansky and Simpson. This *population* dimension, along with the historical, suggests that there are two ways to think about species taxa. We can think about them over time, as historical, *diachronic* entities that originate, change and go extinct. Or we can think about them at a single time, as *synchronic* groups of organisms that are connected or given some sort of structure by biological processes. If so, then evolutionary theory tells us that whatever else species taxa are, they have two dimensions – diachronic and synchronic. An adequate theoretical species concept must reflect that fact.

These theoretical commitments are reflected in the hierarchical pluralisms advocated by Mayden and de Queiroz. As we saw in chapter 4, Mayden asks which species concept has the most theoretical significance, and then argues for the *evolutionary species concept*, that conceives species as "a lineage of ancestor-descendent populations" with a distinctive "evolutionary role," a distinctive "evolutionary tendency," and a distinctive "historical fate." This theoretical species concept satisfies the historical component implicit in evolutionary theory by virtue of being a historical lineage connected by ancestor–descendent relations. And given that this is a lineage of *populations*, it also satisfies the synchronic component of species as groups of organisms at particular times.

But this definition also raises many questions. What precisely are these populations and how do they have ancestor–descendent relations? What precisely is a lineage? What are these distinctive evolutionary roles and tendencies? What is a historical fate? The answers to these questions, whatever they are, will presumably clarify and develop the *evolutionary species concept*. In Whewellian terms, these questions are asking for an explication of the concept. In part, the answers will come from evolutionary theory, broadly construed. Population genetics, for instance, has much to say about what a population is. Ecology might well have something to say about evolutionary roles in terms of ecological niches. Developmental evolution might well have something to say about evolutionary tendencies in terms of developmental constraints. Theories of macroevolution may have implications here as well, answering questions about historical fates. What this all suggests is, first, that there is potentially great theoretical significance to this species concept insofar as it engages the evolutionary theoretical framework in many ways. Second, because the ways the *evolutionary species concept* gets explicated relative to factors such as evolutionary roles, tendencies and fates are not specified, it is vague in crucial and important ways. It does not specify particular kinds of populations or lineages. This suggests that it can apply to population lineages throughout biodiversity, sexual and asexual, branching or reticulating. Mayden sees this as a virtue of the *evolutionary species concept*, making it universal: "One concept, the ESC demands only that speciation and evolution are natural processes involving lineages that maintain cohesion and have unique identities" (Mayden 1997: 416). This concept leaves open the different ways for a population lineage to play an evolutionary role, maintain cohesion, and have an identity.

As we also saw in the previous chapter, Kevin de Queiroz advocates a similar theoretical concept in his hierarchical pluralism, dubbing it the *general population lineage* concept. On this approach species are historically extended segments of "metapopulation lineages" (de Queiroz 2005: 6601). As with Mayden's approach, there are questions about the nature of lineages, populations and how they are segmented. According to de Queiroz a lineage is "a series of entities forming a single line of direct ancestry and descent" (de Queiroz 1999: 50). He then distinguishes different kinds of biological lineages:

> Biological entities at several different organizational levels form lineages. Thus, biologists speak of gene lineages, organelle lineages, cell lineages, organism lineages ... and population lineages. Because entities that form lineages often make up, or are made up of, entities at different organizational levels, the same is also true of the lineages themselves. An organism lineage, for example, is (often) made up of multiple cell lineages, and multiple organism lineages make up a population lineage. (de Queiroz 1999: 50)

Species lineages are a particular kind of lineages, occurring above the organismic level, at the population level, which can come in either biparental (sexual) or uniparental varieties (de Queiroz 1999: 52). Populations occur at different levels as well, with the corresponding lineages:

> The population level is really a continuum of levels. Lineages at lower levels in this continuum (e.g., demes or deme lineages) often separate and reunite over relatively brief intervals. Toward the other end of the continuum, lineage separation is more enduring and can even be permanent. (de Queiroz 1999: 53)

We can in principle identify as populations small groups of organisms that interact and share a niche – demes, to subspecies that occupy a substantial geographic area, or higher to species that occupy entire continents, but yet still interbreed if given the opportunity.

Each population level will have its own lineage, and ways of segmenting the lineage at that level. The species level of population, lineage and segment will have its own distinctive features:

> Under the lineage concept of species, species are not equivalent to entire population lineages, but rather to segments of such lineages. Just as a cell lineage is made up of a series of cells and an organism lineage of a series of organisms, a species (population) lineage is made up of a series of species. Not just any lineage segment qualifies as a species, however. Instead, a species corresponds with a lineage segment bounded by certain critical events. (de Queiroz 1999: 53)

But de Queiroz declines to give the specific ways for identifying species population lineages segments, in order to make the concept more widely applicable across diversity:

> This definition describes a very general conceptualization of the species category in that it explains the basic nature of species without specifying either the causal processes responsible for their existence or the operational criteria used to recognize them in practice. It is this deliberate agnosticism with regard to causal processes and operational criteria that allows the concept of species just described to encompass virtually all modern views on species, and for this reason, I have called it the *general lineage concept of species*. (de Queiroz 1999: 53)

The idea is that in order to accommodate all kinds of organisms, the primary theoretical concept must *not* specify which processes are responsible for the populations that form the lineages and segment them into species. This vagueness in the *general population lineage* concept, like the vagueness in Mayden's *Evolutionary Species Concept*, is a virtue.

Primary theoretical concepts, according to hierarchical pluralism, are evaluated in terms of theoretical significance and universality. Does a particular concept have the right theoretical significance to function within the greater evolutionary theoretical framework? And does it have the universality required to apply across biodiversity? Whether the theoretical concepts of Mayden and de Queiroz are theoretically adequate and universal, is not to be decided by philosophers, but rather in their development or "explication," in light of theory and application to biodiversity. In other words, the evaluation of these concepts is to be determined by their success in scientific practice. Nonetheless, there are some tentative conclusions we can draw. First, evolutionary theory tells us that species have a history and are related by ancestor–descendent relations. Second, there is also a dimension to species at a particular time – whether it be in terms of population or metapopulations, or demes. A concept that does not recognize both dimensions will be theoretically inadequate. Third, vagueness is a virtue. A satisfactory theoretical concept cannot be so specific about the processes that produce the particular kinds of segments in population lineages that it does not apply across biodiversity. It cannot specify, for instance, the reproductive isolation found in sexually reproducing species, because that would make it inapplicable to large portions of biodiversity. Finally, while a theoretical species concept need not itself be operational, there must be ways to connect it to nature. We need to be able to tell when we have identified and individuated the relevant segments of population lineages. For both Mayden and de Queiroz this is accomplished through operational concepts.

OPERATIONAL CONCEPTS

Whewell argued that scientific concepts apply to the world by colligating facts. These facts can be directly observable, such as the phases of the moon, or less directly observable, such as Kepler's laws of planetary motion. In our application of Whewell's model to hierarchical pluralism, theoretical concepts colligate facts and become more consilient through explication. But, as the hierarchical model recognizes, theoretical concepts require operational concepts. Theoretical concepts colligate facts *through* operational concepts. Among these operational concepts, which connect theoretical concepts to observation, are many of the concepts described in the previous chapter, from the morphological and genetic concepts, to the interbreeding biological species concept of Ernst Mayr, the concepts based on gene pools and reproductive communities, specific mate recognition systems, and the concepts based on adaptive zones and ecological functioning. Each of *these* species concepts represents a potential way to connect the theoretical concept to nature, with all its diversity and variability.

These secondary operational concepts, in contrast to the primary theoretical concepts, are not themselves evaluated in terms of their consilience, but in terms of *operationality* and *theoretical relevance*. On the operationality criterion, we evaluate species concepts in terms of the ease of observation and application to the phenomena. How easily can a concept be used to identify and individuate species taxa? On the theoretical relevance criterion, we evaluate in terms of relevance to the primary theoretical concept and its functioning within the overarching theoretical framework. Does an operational species concept pick out those features of biodiversity that are relevant to what species taxa are, given the primary theoretical concept? We can see how operational concepts work with many of the species concepts employed by modern systematists.

The facts colligated by operational concepts range on a continuum from the most directly observable to partially observable to entirely unobservable. At each level, there is a set of theoretically based assumptions that guide the application of operational principles, telling us what similarities and processes are relevant, and how they are relevant. The most directly observational facts to be colligated include those about patterns of similarities among organisms. Morphological similarities and differences are normally directly observable, and are therefore highly operational. We simply see shapes, colors, size, correlation of parts and so on, and classify on that basis. But obviously, not all morphological similarities are relevant to species groupings. That is determined by what we know about species lineages – what our theoretical framework tells us about species lineages and the processes that result in morphological similarity and difference. This framework tells us, first, that offspring typically resemble their parents. But, especially among sexual organisms, there are typically morphological differences among parents and offspring – differences in size, form, coloration and so on. Some of these differences are due to sexual dimorphism and developmental stages. Among birds, for instance, interbreeding males and females often have striking differences in coloration. Many insects go through extremely different developmental stages. Other differences are due to the effect of the environment, such as those correlating the size of plants and elevation. These differences are all irrelevant to species groupings. We don't group male songbirds into separate species taxa from their less colorful mates. A morphological or phenetic concept therefore requires theoretical guidance to indicate which similarities are relevant to species groupings. That requires reference back to the primary theoretical concept and demands an answer to the question: which similarities are indicative of membership in a particular segment of a population lineage?

We must answer similar questions when using genetic similarities to group organisms into species. Parents and offspring within a lineage tend to have similar, but not identical, genotypes, especially in sexually reproducing organisms. And within a lineage, genotypes tend to change over time to varying degrees. Similarly, genotypes vary within populations at a time, in regular and predictable ways. Some genetic similarities are therefore relevant to the identification of population lineages, and can serve as operational guidance, while some genetic similarities are not relevant. Genetic similarity can therefore serve as an operational criterion to identify and individuate lineages – given suitable theoretical assumptions. This implies that morphological and genetic species concepts can have operational value for the application of a theoretical species concept based on the idea of a population lineage. This is true even of past population lineages. As we are able to gather morphological and genetic information from fossils, we have more information to reconstruct the population lineages of the past.

Species concepts based on various processes can be of operational value as well, depending on theoretical relevance. Among sexual species, for instance, the ability to interbreed is relevant to the segmentation of population lineages. Two individual organisms may have very different morphologies, but if they can interbreed and produce fertile offspring we would likely group them into the same species. On the other hand, in plants there is substantial hybridization among what are typically identified as different species. And among asexual organisms, this operational principle is irrelevant. Consequently, interbreeding is operationally relevant to the identification and individuation of *some* species-level segments of population lineages. The divergence of these lineages is facilitated by the cessation of interbreeding, or hindered by continued interbreeding. This is, of course, the idea behind Ernst Mayr's biological species concept. Notice that there is a range of observability here. Interbreeding, or lack of interbreeding, can sometimes be observed directly, in sympatric populations. But sometimes it must be

137

inferred from differences in morphology, chromosomes and behavior, and particular mate recognition systems.

We might focus on other processes in the origination, maintenance and segmentation of population lineages. According to ecological species concepts, we can identify and individuate species by reference to adaptive zones. A species population lineage may have cohesion by virtue of the selective forces of its environment. Here the operation of natural selection will play a role in identifying particular populations at a particular time as well as over time. Perhaps natural selection will be stabilizing within a population, preventing change. Or it might promote change over time. In these cases, the theoretical framework, population genetics and population ecology can help us determine the relevance of natural selection to the identification and individuation of segments of population lineages. Similarly, geographic factors might, given the right theoretical support, be relevant to species determinations. Reproductive barriers from geographic isolation are relevant to distinguishing species as population lineages, although not always fully determinative.

What is important here is that these operational concepts are all *potentially* relevant to the identification and individuation of species taxa in terms of population lineages. If and how they are relevant is to be determined by the theoretical framework. What does the evolutionary framework, as it is currently understood, imply about the relevance of particular morphological and genetic similarity, interbreeding and mate recognition systems, ecological functioning, and geographic distribution? Second, these concepts are useful insofar as they are operational. How can we make the required determinations to identify and individuate species taxa? Can we get the data required? Genetic similarity has not been widely applicable to fossils, so it has not been of general use. But recent technological developments are changing that. Interbreeding tendencies are observable in sympatric populations, but not in the allopatric populations that don't overlap. But observation of physiology and knowledge of behavior can provide some operational guidance about interbreeding in these populations. Some physiological differences in sex organs simply make reproduction impossible, while differences in mating behaviors can make it unlikely.

Whether an operational concept is theoretically relevant depends on what our best theories tell us about the segmentation of population lineages and evolutionary change. And whether an operational concept is truly operational is an empirical matter. It depends on facts about the natural world. Whether a genetic species criterion applies to fossils depends on the availability of satisfactory genetic material. This depend- ence on both the theoretical framework and empirical facts implies that the application of operational concepts will be complex and difficult. We cannot say beforehand which operational concepts are going to be relevant, and our conclusions about relevance will likely change along with changes in the theoretical framework. With new technology, the operationality of particular concepts will change as well. This all sug- gests that, with operational concepts, we should embrace a *principle of proliferation*. The more operational guidance we have in identifying segments of population lineages, the better. So instead of aiming here for consilience – unification under a single concept – we should aim for the multiplication or proliferation of species concepts. The more the merrier!

CORRESPONDENCE RULES AND THE SPECIES PROBLEM

So far I have followed Mayden in his terminology, in distinguishing theo- retical *concepts* from operational *concepts*. As long as we keep in mind the differences in the functioning of these two kinds of concepts, this poses no problem. But because theoretical and operational concepts function differently, we might well use different terms. Mayden some- times used the term "guideline" instead of "concept" when referring to operational concepts. Similarly de Queiroz sometimes used the term "criteria" for the operational concepts, in contrast to the "definitions" provided by theoretical concepts. There is an important and not merely semantic point here that echoes another philosophical dispute about sci- entific concepts. A brief look at that dispute may help us clarify the solu- tion to the species problem outlined here.

Early in the twentieth century there was a debate about how to define scientific concepts in physics such as *length* and *mass*. The physicist P. W. Bridgman proposed that we should define these concepts in terms of the *operations* we use to measure them. *Mass*, for instance, would be defined in terms of the ways of measuring mass – the resistance to accel- eration, or the operation of gravity. Bridgman's proposal that operations give definitions became known as "operationalism," and came to be applied to the philosophical problem of how to connect theoretical laws that contain only non-observational terms to observation. How can we connect laws about unobservable particles, for instance, to the empirical

regularities we observe in nature? The philosopher of science Rudolf
Carnap explained:

> Our theoretical laws deal exclusively with the behavior of molecules,
> which cannot be seen. How, therefore, can we deduce from such laws a
> law about observable properties such as the pressure or temperature of
> a gas or properties of sound waves that pass through the gas? The theo-
> retical laws contain only theoretical terms. What we seek are empirical
> laws containing observable terms. Obviously, such laws cannot be derived
> without having something else given in addition to the theoretical laws ...
> That something else that must be given is this: a set of rules connecting
> the theoretical terms with the observable terms. (Carnap 1966: 233)

Carnap called these rules "correspondence rules," but saw them as
equivalent to P. W. Bridgman's "operational rules" and N. R. Campbell's
"dictionary." What is significant in Carnap's proposal is that these correspondence rules connecting theoretical concepts to observation are
not really concepts in the usual theoretical sense. This is clear in his
rejection of the view (disagreeing with Bridgman) that operational rules
can provide definitions: "There is a temptation at times to think that
the set of rules provides a means for defining theoretical terms, whereas
just the opposite is really true" (Carnap 1966: 234). Nor can correspondence rules function *as* definitions: "What we call these rules is, of course,
only a terminological question; we should be cautious and not speak of
them as definitions. They are not definitions in any strict sense" (Carnap
1966: 236). Rather, the definitions give operational guidance, telling us
what operations are relevant. What is important here is that those concepts that function operationally are different from those that function
theoretically. They tell us how to observe a thing, not what sort of a
thing it is.

We can apply Carnap's insight here to the species problem. As argued
by Mayden and de Queiroz, some species concepts are theoretical. They
tell us how to conceive species. They define species taxa and constitute the species category. But some species concepts are operational.
They tell us how to identify and individuate the groups that are properly species *given a particular theoretical concept.* But these so-called
operational concepts are really *rules* that help us to determine if a group
of organisms satisfies the demands of the theoretical concept. Carnap
called these so-called operational concepts "correspondence rules."
By using this terminology, we could avoid the confusion of definitions
with operations. He thought this solved a general problem in science,

the tendency of philosophers to ask scientists for definitions of scientific concepts in familiar, non-theoretical terms.

> They want the physicist to tell them just what he means by "electricity", "magnetism", "gravity", "a molecule". If the physicist explains them in theoretical terms, the philosopher may be disappointed. "That is not what I meant at all", he will say. "I want you to tell me, in ordinary language, what those terms mean." (Carnap 1966: 234)

The problem here is that the scientist is being asked for something he or she cannot give – a definition in operational rather than theoretical terms. Each of these concepts has satisfactory definitions, but they are in terms of the theoretical framework. That is the proper source for definitions – telling us how to interpret these concepts – not the operations to measure or identify the things that satisfy them. Carnap concluded:

> The answer is that a physicist can describe the behavior of an electron only by stating theoretical laws, and these laws contain only theoretical terms. They described the field produced by an electron, the reaction of an electron to a field, and so on … There is no way that a theoretical concept can be defined in terms of observables. We must, therefore, resign ourselves to the fact that definitions of the kind that can be supplied for observable terms cannot be formulated for theoretical terms. (Carnap 1966: 235)

Carnap's analysis here is relevant to the species problem in two ways. First is the role of theoretical frameworks in the interpretation of scientific concepts. The theoretical framework is fundamental in the interpretation of species concepts, while the operational rules are subservient to the theoretical. In chapter 7, we will look more closely at this role of theory in conceptual interpretation and change. Second is the proposal that we think about operations as rules rather than concepts. What we might call operational *concepts* are really *rules* for connecting theoretical concepts to observation. By using this terminology, we are better representing and reflecting the division of conceptual labor.

CONCLUSION

In this chapter we first examined the skeptical solutions to the species problem that deny there are any species things, or that there is a single kind of species thing. These skeptical solutions have their costs, threatening the assumption that species are the fundamental units of evolution and classification. I then suggested that, if we wish to preserve this significance of the species category, we should continue the search for a single, satisfactory way of conceiving species. This solution, I argued, is to be found in the hierarchical pluralism developed by Mayden and de Queiroz, which distinguishes the roles of theoretical and operational concepts. This division of conceptual labor solution suggests first that we should evaluate the different kinds of concepts on different grounds. A theoretical concept should be evaluated in terms of its theoretical significance relative to an overarching theoretical framework. For species concepts this means significance relative to evolutionary theory. A theoretical concept should also unify phenomena insofar as unification is possible. A theoretical species concept does this by virtue of its universality – applying across biodiversity as much as possible, to sexual and asexual species organisms, vertebrates, invertebrates, bacteria and fungi. By contrast, operational concepts must be theoretically relevant and operational. This means that they should facilitate the identification of species taxa through factors determined to be relevant by the theoretical concept, that in turn has the most significance relative to evolutionary theory. Because the operational goal is best served by reference to many factors, morphology, reproduction, etc., we should adopt a *principle of proliferation* relative to operational concepts. The more ways we have of identifying and individuating species taxa – segments of population lineages – the better. This division of conceptual labor solution is therefore theoretically monistic and operationally pluralistic.

Whether or not this solution turns out to really solve the species problem remains to be seen. We need to determine how universal the theoretical concept based on the idea of segments of a population lineage can be. With sexually reproducing vertebrates, there are few obvious worries. But with hybridizing plants and asexually reproducing invertebrates, bacteria and fungi there may be worries about the application of a general lineage concept as developed by Mayden and de Queiroz. We may ultimately be forced to follow Ereshefsky in postulating more than one kind of lineage. Or we may conclude that some organisms simply do not form species taxa. The point here is that the adequacy of any theoretical concept is to be determined by its application in scientific practice to phenomena, facilitated by the appropriate set of correspondence rules. Even if, as de Queiroz has argued, the general lineage concept has been widely accepted, there has not been conscious and systematic application of that concept across all biodiversity. If there had been, we would not be worrying about the species problem.

Nonetheless, this hierarchical framework based on the division of conceptual labor has its own virtues, independent of the adequacy of particular theoretical concepts. First, it has the potential to preserve the significance of species in evolutionary theory – whatever that significance may be. If, on our best theories, the species category loses its significance, then the hierarchical approach will reflect that fact. Or if the significance of the species category changes relative to the theoretical framework, the hierarchical approach will reflect that as well. And if there is a need for a single theoretical concept, as there seems to be now, then that will be reflected in the hierarchy. If we adopt Whewell's approach, we can also be prepared to look at how the species concept as population lineage gets explicated in the debates among biologists and systematists. How do the various debates explicate – develop, clarify and refine – the idea of a population lineage?

Second, this solution to the species problem is consistent with actual practice. It recognizes the many operational concepts – *correspondence rules* – employed by biologists that vary by organism studied and theoretical interest. On the hierarchical division of conceptual labor solution, we would expect that morphologists would be looking at morphology, ethologists would be looking at behavior, molecular biologists would be looking at molecules, ecologists would be looking at adaptive niches, and developmental biologists would be looking at developmental programs and constraints. All of these factors are *potentially* relevant to understanding the nature of species as segments of population lineages – even if that is not the immediate goal of the investigation. In other words, this solution explains the proliferation of operational approaches. The more we know about biodiversity, the more operational options we find, the more *correspondence rules* we discover.

The third virtue of this solution is that it explains the species problem at a more fundamental level. We have the species problem because we have not had a clear understanding of the conceptual framework in which the species concepts function. We have not explicitly recognized the division of conceptual labor. The evaluation of species concepts has typically been relative to a single set of criteria taken to apply to all concepts – theoretical significance, universality and operationality. If there is a division of conceptual labor, this evaluative framework *guarantees* a species problem. We have treated the biological and morphological concepts as competitors, rather than as complements. The biological species concept is not a competitor to the population lineage concept, but a way to recognize some kinds of population lineages. The bottom line is that we have a species problem partly because we have conflated different kinds of species concepts, some of which are operational, and not concepts in the same way as others. They are instead criteria, guidelines or *correspondence rules* for connecting theoretical concepts to empirical phenomena.

We can also see how we came to have this problem. The idea of a species antedated evolutionary theory by two millennia in the philosophies of Plato and Aristotle, and their use of the term *eidos*. Its history extends through the medieval scholastics, Renaissance herbalists and pre-Darwinian naturalists. There is therefore a long tradition of thinking about species that is uninformed by modern evolutionary theory. Like the homologies we see in organisms that were inherited from their ancestors, there are ideas about species we have inherited from pre-evolutionary thinking. These homologous ways of thinking about species are embodied in the Linnaean hierarchy, the emphasis on morphology and the idea that we can *define* species taxa. The taxonomic tradition has in effect worked in opposition to the Darwinian revolution and its implications for the species category. The bottom line is that we have a species problem partly because we have a history of thinking about species that is not theoretically grounded in evolution.

There are two additional topics that are worth introducing here, and that will be explored in the following chapters. First, because the species problem and its solution are found in the conceptual framework, we can understand that problem better if we can get a better understanding of how the framework functions. But we can also better understand the species problem if we get a better understanding of how *individual* concepts work. How do species concepts get their meaning and how do they change? In chapter 7, we will look more closely at these questions. There are also a set of philosophical issues lurking here related to metaphysics, and the "ontology" of evolution. How should we think about species taxa at a more basic level? Are they *sets* of organisms with members, as traditionally assumed, or are they *individual* things with parts? This is the topic of the next chapter.

6

Species and the metaphysics of evolution

INTRODUCTION: SPECIES CONCEPTS AND METAPHYSICS

In the previous chapter, I sketched out a potential solution to the species problem based on an idea implicit in the hierarchical pluralism of Mayden and de Queiroz, the division of conceptual labor. This solution distinguishes theoretical concepts that tell us what species things are, from operational concepts, or more accurately *correspondence rules*, that tell us how to identify and individuate species taxa. I also tentatively endorsed the basic idea behind the theoretical concepts proposed by Mayden and de Queiroz that treats species as segments of population lineages. This idea reflects the basic evolutionary assumption that species taxa have two dimensions, synchronic and diachronic. They exist at a particular time as population of organisms, and over time as a lineage of ancestors and descendants. But even if we were to accept as unproblematic this more general claim, along with the specific conception of species as segments of population lineages, there would still be questions about the nature of species taxa. There is still more to the species debate.

There is also a *metaphysical*, or more specifically, an *ontological* question: Given that species taxa have the features we think they have, what fundamental kinds of things are they? This question is asking us to think about species at a more general, and more fundamental level. In this chapter, we will be looking at this question, how it has been answered and why, and what the implications are for each metaphysical (or ontological) stance. One promising answer is associated with Michael Ghiselin and David Hull, that species are spatio-temporally restricted *individuals*. This *species-as-individuals* thesis, seems to cohere better with evolutionary theory, and have greater heuristic value in the further development of evolutionary thinking than the alternative stance, that

145

species are *sets* of organisms. But before we look at these competing metaphysical stances, it will be useful to see what is at stake, and how we should think about the metaphysics in general.

In popular usage, the term *metaphysics* has a problematic history, sometimes referring to what lies beyond the natural world and what we might describe as "paranormal." But even if we don't think of seances and palm reading, this term is hard to pin down. As is well known, its origin is in Andronicus' compilation of Aristotle's work, where the *Metaphysics* was merely the set of writings that came "after the *Physics*" – "te meta ta phusika." When Aristotle described what he was doing in these writings, he used terms usually translated as "first philosophy," "first science," and "theology." Topics included "being as such," and "first causes." The reasons for the placement of these writings after the *Physics* are unknown. It may be that these works were just regarded as more difficult than those of the *Physics* and thus should be studied after the *Physics*.

Contemporary metaphysics, as taught in universities, often seems to consist of a grab bag of unconnected questions about topics that don't fit well into other philosophical categories. A textbook might contain topics on determinism and the possibility of free will, necessity and contingency, properties, objects, time and the existence of God. There is, however, a conception of metaphysics that sees it as a more coherent project, an "exploration of the most general features of the world." Simon Blackburn explains:

> Metaphysics is the exploration of the most general features of the world. We conceive of the world about us in various highly general ways. It is orderly, and structured in space and time; it contains matter and minds, things and properties of things, necessity, events, causation, creation, change, values, facts and states of affairs. Metaphysics sees to understand these features of the world better. It aims at a large-scale investigation of the way things hang together. (Blackburn 2002: 61)

Metaphysics, in this sense, does not go beyond the world we experience, but addresses it at a higher level of generality. It doesn't ask what specific things are in the world, but *given* the specific things we find in the world, is there some more coherent and general way to think about them? Descartes was engaged in this kind of a project when he argued that there were two kinds of substances – an extended material substance and a non-extended thinking substance.

Related to this way of thinking about metaphysics, as pertaining to the most general features of the world, is a naturalistic approach that tells us to proceed on the basis of what science tells us about the world. In metaphysical investigations, we start with science and then proceed to a more general account of nature. This *naturalistic* metaphysics, in its most comprehensive form, looks to science to answer *all* questions about the most general features of the world. One advantage of this global naturalism is that it gives us a way to proceed about even the most difficult metaphysical questions: we proceed scientifically. There is some resistance to such a naturalistic metaphysics among philosophers. Some see this as nothing more than "science envy," a case of philosophers looking to capture some of the prestige of science. (Blackburn 2002: 76). We need not adopt the global project though, to see a connection between metaphysics and science. Each science seems to have its own metaphysical issues. When Whewell asked us to think about Newtonion gravity more generally and fundamentally as a force, he was asking us to think about the world in more general terms. Given what we observe about gravity, its causal role in planetary motion, the descent of stones on Earth and the tides, what is it at a more fundamental basic level? The answer Newton gave was that it was a *force*, a more basic, fundamental kind of thing.

We can think of species in this same way. Given what we know about species taxa, their diachronic and synchronic dimensions, the role of interbreeding, the operation of natural selection in an environment and the patterns of morphological and genetic similarities, what are they at a more general and fundamental level? There have been two main answers to this question. The first tells us that species are *sets* (or classes) of organisms, and organisms are *members* of the set (or class). The second tells us that species are concrete *individuals*, and that organisms are its *parts*. In this chapter, we explore these two ways of conceiving species at a more general level. But first, we need to be more clear about what we are doing in pursuing the metaphysical questions.

The most prominent advocate of naturalistic metaphysics in the twentieth century, W. V. O. Quine, asked us to think about metaphysics in terms of the *ontological* presuppositions of our philosophical and scientific theories, and the language we use to express them. These ontological commitments, about what fundamental kinds of things exist, are just as much a part of science as are the more directly empirical claims about observation and theory. Quine saw this as a continuum:

> Within natural science there is a continuum of gradations, from the statements which report observations to those which reflect basic features say

of quantum theory or the theory of relativity. The view which I end up
with ... is that statements of ontology or even of mathematics and logic
form a continuation of this continuum, a continuation which is perhaps
yet more remote from observation than are the central principles of quantum theory or relativity. The differences here are in my view differences
only in degree and not in kind. (Quine 1976: 211)

The most important idea here is that fundamental metaphysical
questions about ontological commitments are not separate, "external"
questions. They are "internal" and a proper part of science. If so, then
questions about whether species taxa are ultimately sets or individuals
are legitimately a part of biology and are to be resolved from within
biology – even if these are largely philosophical issues. If so, we need to
know the biology; we need to take it seriously; and we need to incorporate what we know about it into our metaphysics. This is as true for
the philosophers who think about species as it is for the biologists.

One more clarification is required before we turn to the main question here. Metaphysics can be either descriptive, telling us what basic
sorts of things there are, or prescriptive, telling us how we *should* conceive things. The prescriptive approach is often revisionary, telling us
that we should reject an old metaphysical commitment and adopt a
new. These two approaches are not independent, however, revisionary
accounts often depends on the descriptive, as Blackburn explains:

> Metaphysics may be a purely descriptive enterprise. Or, it may be that there
> is reason for revision: the ways we think about things do not hang together,
> and some categories are more trustworthy than others. Revisionary meta-
> physics then seeks to change our ways of thought in directions it finds
> necessary. The distinction between revisionary and descriptive metaphys-
> ics is not sharp, for it is out of the descriptions that the need for revision
> allegedly arises. (Blackburn 2002: 61)

I follow Blackburn here in that my project will be both descriptive and
prescriptive. It will start with an understanding of the metaphysical presuppositions of evolutionary theory. We begin with what that theory tells
us about the basic, fundamental nature of species taxa. But it also contains elements of revision. We can refine and develop a metaphysics of
evolution so that it better serves the theoretical goals and reflects empirical facts about the world. But before we turn to what evolutionary theory tells us about the fundamental nature of species, we should look to
what philosophers have told us about the metaphysical status of species,
and how it has gone wrong.

In previous chapters I have been critical of the Essentialism Story. According to this story, before Darwin species taxa were predominantly conceived as *natural kinds* with *essences* – a set of essential properties that tell us what kind of a thing an organism is. (Or on the alternative formulation, species are natural kinds with a set of individually necessary and jointly sufficient properties.) As we saw in the work of historians of science such as Mary Winsor, and historically inclined philosophers of science such as Phillip Sloan, there was no essentialist consensus in pre-Darwinian biology. This was borne out as we traced the history of thinking about species in previous chapters from Aristotle to modern biologists and systematists. But while the Essentialism Story is wrong about the views of naturalists, biologists and systematists, there is a powerful essentialist tradition in philosophy.

A central component of this Essentialism Story is that species taxa are *natural kinds*. Recall the words of Elliott Sober from chapter 2:

> *Essentialism* is a standard philosophical view about natural kinds. It holds that each natural kind can be defined in terms of properties that are possessed by all and only members of that kind. All gold has atomic number 79, and only gold has that atomic number. It is true, as well, that all gold objects have mass, but *having mass* is not a property unique to gold. A natural kind is to be characterized by a property that is both necessary and sufficient for membership. (Sober 2000: 148)

While Sober does not endorse this essentialism with regard to species (as we shall see later in this chapter), many philosophers have endorsed it, treating biological species as natural kinds with essences. What then are *natural kinds*? Or more fundamentally, what are *kinds*?

Kinds are typically treated as *sets* (or classes) of things, based on the possession of a common property. For instance the set of all red things can form a kind on the basis of the property of being red. Joseph Laporte describes kinds in this way: "for any property, there is a corresponding kind the essential mark of which is to possess that property ... For example, to the property of being red there corresponds redkind" (LaPorte 2004: 15). Here there is a trivial sense in the claim that being red is essential, since it is just that property that was used to constitute the kind. There is something arbitrary in this sort of essence and associated kind. But *natural* kinds are different. They are commonly taken to be independent of human interests, language and epistemic considerations, and thereby reflect true divisions of the world. Paradigmatic natural kinds such as *water, electron* and *planet* are natural kinds because they are out there in the natural world, not just in our way of thinking about the world. When we divide the world into these sorts of things, we are, as the oft-quoted saying goes, "cutting nature at its joints."

These natural kinds are sometimes contrasted with conventional or artificial kinds. *Conventional kinds* are classes of things that are what they are merely by human convention. A particular piece of paper is a dollar bill because it is conventional to treat it as such, and *only* because we treat as such. We might classify french fries as a vegetable because it is conventionally treated that way on menus. *Artificial kinds*, by contrast, might be based on convention, but might not. There may be no convention that treats all red things as a kind, but we can still class them that way, and for whatever reason. We can similarly postulate such arbitrary artificial kinds as "named on Tuesday" – the set of machines, animals, humans, planets, etc., that were named on some Tuesday or other (LaPorte 2004: 19).

In recent literature, natural kinds have come to be closely related to the practice of science. Natural kinds are just what science studies, as Alexander Bird explains:

> Scientific disciplines divide the particulars they study into kinds and theorize about those kinds. To say that a kind is natural is to say that it corresponds to a grouping or ordering that does not depend on humans. We tend to assume that science is successful in revealing these kinds; it is a corollary of scientific realism that when all goes well the classifications and taxonomies employed by science correspond to the real kinds of nature. (Bird and Tobin 2008: 1)

Those who adopt this approach typically claim that natural kinds support inductive inference; participate in the laws of nature; form a hierarchy and are categorically distinct (Bird and Tobin 2008: 6). Because *electron* is a natural kind, for instance, we can generalize about electrons. There are laws governing electrons, electrons are part of a hierarchy of particles, and they are distinct from other particles. More generally, natural kinds, as opposed to conventional or artificial kinds, are taken to play a role in scientific explanation. We can explain some phenomena, for instance, in terms of the transfer of electrons.

Typically natural kinds are taken to be kinds because of their *intrinsic* properties (Bird and Tobin 2008: 19). Intrinsic properties are usually contrasted with *extrinsic* properties that depend in some way or other on some external state of affairs. Most obviously, size, shape, and constitution are taken to be intrinsic. A thing has particular properties of shape and size, and is made of particular substance or component parts independently of other objects or states of affairs. By contrast, something has particular properties of being *larger than, above, descended from*, and *sister of*, only because of some relation to an external thing or state of affairs. An electron is a natural kind then, because of its negative charge, and not because of its spatio-temporal location, or relation to other particles. A substance is of the kind *copper*, because of its atomic number. Another substance is of the kind *water* because of its particular molecular composition. What properties are and how they work is complex and beyond the project here. (See Mellor and Oliver 1997.) But we can note for now that the requirement that properties be intrinsic, rather than extrinsic or relational, may be too restrictive. As we shall soon see, some philosophers want to explicitly incorporate relational properties into an account of essences.

The association of natural kinds with science has figured centrally in one of the most influential debates about natural kinds of the last century. Saul Kripke (1972) and Hilary Putnam (1990) developed essentialist views about natural kinds based on a causal semantics. According to the Kripke-Putnam approach, essences get established through the application of natural kind terms to things in nature. We point to, and name, a particular substance such as *water*. Then science proceeds to determine the essential properties of that substance. Chemistry discovers that water is composed of two hydrogen atoms and one oxygen atom, and it could not be otherwise. Anything that had a different atomic structure would not, and cannot be water. For Putnam the reference of these natural kind terms is established through a division of linguistic labor. Because chemists are the experts about things like water, when they collectively point to it and investigate its nature in light of some scientific theory, their conclusions set the extension of the term *water*. And once the extension of the term is set, we can discover necessary facts about which properties are essential. Chemistry tells us that the molecular structure of water is an essential property, but its transparency and taste are not.

There has been a great deal of philosophical discussion about the causal theory of reference and its associated views about natural kinds and essences. We shall return to some worries about it in the next chapter. But for purposes here, what is most important is that Kripke and Putnam are not just interested in chemical kinds, but think they have also given a satisfactory account of biological kinds. They treat species names as natural kind terms, implying a sort of species essentialism. According to this version of essentialism, naturalists, biologists and systematists apply terms such as *elm* and *tiger* to objects in the world. The reference of these terms thus gets set, and science proceeds to find out what necessary properties each of these objects has that makes them the kind of thing they are – what makes an elm an elm, and what makes a tiger a tiger. Like water, these essences are usually taken to be found in the microstructure, the genetic structure. David B. Kitts and David J. Kitts follow Putnam here:

> The property which all the organisms of a species share and which ultimately accounts for the facts that they cannot be parts or members of any other is not some manifest property such as the pigmentation of a feather. It is an underlying trait. Putnam is not far from the mark concerning the essential nature of lemons when he says, "What the essential nature is is not a matter of language analysis but of scientific theory construction; today we would say it was chromosome structure, in the case of lemons, and being a proton-donor in the case of acids." (Kitts and Kitts 1979: 617–618)

Kitts and Kitts conclude: "Biologists search for the underlying trait which explains the necessary relationship between an organism and its species in the genetic structure of the organism" (Kitts and Kitts 1979: 618).

Similarly, and more recently, Michael Devitt has argued for a version of essentialism along these lines, claiming that "Linnaean taxa have essences that are, at least partly, intrinsic underlying properties" (Devitt 2008: 346). The "Linnaean taxa" that most interests Devitt are species taxa. Like Kitts and Kitts, he thinks the essences of species taxa are primarily (but not exclusively) to be found among genetic properties.

> In sexual organisms the intrinsic underlying properties in question are to be found among the properties of zygotes; in asexual ones, among those of propagules and the like. For most organisms the essential intrinsic properties are probably largely, although not entirely, genetic. Sometimes these properties may not be genetic at all but in "the architecture of chromosomes," "developmental programs," or whatever ... For convenience, I shall often write as if the essential intrinsic properties were simply genetic but I emphasize that my Essentialism is not committed to this. (Devitt 2008: 347)

Devitt looks for essences in these properties, because he thinks that is where the explanatory power lies. We make all sorts of generalization about organisms, about their morphology, physiology, behavior and so on, that beg for an explanation (Devitt 2008: 351). These explanations are typically grounded on the genetic properties of organisms.

> Explanations will make some appeal to the environment, but they cannot appeal only to that. There has to be something about the very nature of the group – a group that appears to be a species or taxon of some other sort – that given its environment determines the truth of the generalization. That something is an intrinsic underlying, probably largely genetic, property that is part of the essence of the group. (Devitt 2008: 352)

There are two things to note here. First, Devitt is following a tradition that identifies essential properties with explanation and explanatory power. This is consistent with the view that science studies natural kinds and their essences. After all, one of the main functions of science is to provide explanations. Second, Devitt is only requiring that essential properties be *part* of the explanatory reason an organism has the features it has. The *essence,* on the other hand, is the *sum* of the essential properties. Since an essence is a conjunction of essential properties, this is a *conjunctive* essentialism.

There is an initial plausibility to this sort of essentialism. Surely there is some set of genetic traits associated with a species that is distinctive to that species and has explanatory value in understanding our generalizations about these members of that species. Tigers have stripes, and this is explained by a set of genes and regulatory developmental networks that causes tigers to develop stripes. Similarly, we read in the popular press that while chimpanzees and humans share a high percentage of DNA, there are differences, and these differences are what makes us human. It is our "essence." This view is less plausible though, when we look at what systematists actually do in grouping organisms into species. If a tiger mating pair has an offspring that lacks stripes, systematists do not therefore conclude it is not a tiger. Generally, they will conclude that it is a tiger (*Panthera tigris*) because it was born of two tigers. While this genealogical relation does not explain the presence or absence of stripes in the way the genetic facts might, it is nonetheless *the* determinative fact. The particular traits are relevant but not determinative. After all, there is great variability within the populations that make up a species, and variation over time. Devitt is aware of this problem, but dismisses it:

> *[A]n intrinsic essence does not have to be "neat and tidy."* And, because the intrinsic essence is identified by its causal work, we need not be concerned that the identification will be ad hoc: the essence of the Indian

rhino is the underlying property that does, as a matter of fact, explain its single horn and other phenotypic features. (Devitt 2008: 371)

But even if this rhino does not in fact have a horn, perhaps because of a mutation in a developmental gene network, it is still an Indian rhino. It is not clear how the genetic basis of the rhino horn can be an essential property, and part of the essence of this species of rhino, if an individual organism will be classified as a member of that species without the horn – and without its genetic basis – simply because it was born of two members of that species. This type of essentialism is simply out of step with systematic practice.

PROPERTY CLUSTER KINDS

Devitt seems to present his version of essentialism as a traditional approach. But one of the virtues of traditional property essentialism is that it is "neat and tidy." There is a set of properties that are individually necessary, and that together are sufficient to make a thing the kind of thing it is. That is the essence of the thing. We see this tidiness in essences of things like electrons, where there are a few determinative, essential properties, based on mass and charge. But with biological species, the list of potentially essential properties is problematic. Stripes are distinctive of *Panthera tigris*, but not determinative. As are size, behavior and mating tendencies. One recent essentialist solution to this problem is based on the idea that there is no *single* set of necessary and sufficient properties that all individual organisms must have to be a member of a species. Rather, essences are a *disjunction*, some combination or other, of a set of properties. One individual can have one subset – a "cluster" – of the relevant properties and be a *Panthera tigris*, another individual can have a different subset and also be a tiger. This idea is usually credited to Ludwig Wittgenstein and his analysis of "family resemblance" in his *Philosophical Investigations*. Two members of a family may resemble each other in one respect, while there may be another, different respect in which they resemble other members of the family. There is a cluster of traits present in this family, and members of that family share some subset of these traits, but not necessarily the same subset (Wittgenstein 1968: 31). This idea of a cluster of traits was applied to biological classification by Morton Beckner, shortly after Wittgenstein's formulation. (See Stamos 2003: 124.) It has also been more recently advocated by Richard Boyd (1999) and R. A. Wilson (1999).

154

Wilson lays out the basic framework of this "Homeostatic Property Cluster" (HPC) approach. There are two components, the disjunction of clustered properties and the mechanisms that cause this clustering:

> The basic claim of the HPC view is that natural kind terms are often defined by a cluster of properties, no one or particular n-tuple of which must be possessed by any individual to which the name applies, but some such n-tuple of which must be possess by all such individuals. The properties mentioned in HPC definitions are homeostatic in that there are mechanisms that cause their systematic coinstantiation or clustering. Thus, an individual's possession of any one of these properties significantly increases the probability that this individual will also posses other properties that feature in the definition. (Wilson 1999: 197)

The first component, the *disjunction,* is a consequence of conditional facts about members of a species, such as whether the individual is a male or female, and the particular stage of life. Boyd explains:

> The fact that there is substantial sexual dimorphism in many species and the fact that there are often profound differences between the phenotypic properties of members of the same species at different stages of their life histories (for example, in insect species), together require that we characterize the homeostatic property cluster associated with a biological species as containing lots of conditionally specified dispositional properties for which canonical descriptions might be something like, "if male and in the first molt, P," or "if female and in the aquatic stage, Q." (Boyd 1999: 165)

The second component is the set of causal mechanisms that cause the clustering of features associated with species and the stability or stasis in clustering.

> A variety of homeostatic mechanisms – gene exchange between certain populations and reproductive isolation from others, effects of common selective factors, coadapted gene complexes and other limitations on heritable variation, developmental constraints, the effects of organism-caused features of evolutionary niches, and so on – act to establish the patterns of evolutionary stasis that we recognize as manifestations of biological species. (Boyd 1999: 165)

Because of sexual reproduction and sexual selection, for instance, there is a stable sexual dimorphism. Because of the operation of natural selection in an ecological niche, there is stability in the type and distribution of morphological traits in general. The important idea here is that there are a number of mechanisms that are operating and that give a species a set of relatively stable range, variety and distribution of traits. Each species taxon will have some distinctive set of traits determined by these homeostatic mechanisms.

There is at least one advantage of the homeostatic property cluster approach over the essentialism advocated by the Kitts and Devitt. Its disjunctive nature makes it more plausible given the general variability within populations, sexual dimorphism and stages of development. When identifying an individual bird's species, for instance, the first and most obvious question to be asked is whether it is a male or female. The traits associated with the males of a bird species are typically different from those associated with the females. The property cluster approach recognizes and accommodates this obvious fact. And it tells us what *kinds* of variation to look for in the property clusters, on the basis of what is known about the processes that generate homeostasis.

But there is an obvious problem with both ways of conceiving species as natural kinds with sets of essential traits – whether those sets are conjunctive or disjunctive. Evolutionary theory tells us that species taxa have both synchronic and diachronic dimensions. They consist of a group of organisms at a single time, but they also consist of a group of organisms connected by genealogy over time. Moreover, they have a *temporal structure*: they have a beginning in speciation; they change over time; and have an ending in new speciation or in extinction. The metaphysical question we asked at the beginning of this chapter was: what general, fundamental sorts of thing are species – given what evolutionary theory tells us about them? Evolutionary theory tells us that they are temporally extended things. Species taxa have a history. Neither the traditional property nor the property cluster approach reflects this fact.

One common criticism of traditional property essentialism is based on precisely this worry. Natural kinds are (presumably) timeless and eternal, and the set of essential properties that make an organism a natural kind is also timeless and eternal. But evolutionary change implies a change in the properties of organisms. Hence, species cannot evolve if they are natural kinds. And if they evolve they cannot be natural kinds. (See for instance Brogaard 2004: 224.) Devitt has a response ready, though, arguing that even though the natural kind cannot itself evolve, given that its essential properties are timeless and unchanging, organisms within a lineage can still change – by passing from one kind to another.

> Suppose that S1 and S2 are distinct species, on everyone's view of species, and that S2 evolved from S1 by natural selection. Essentialism requires

that there be an intrinsic essence G1 for S1 and G2 for S2. G1 and G2 will be different but will have a lot in common. This picture is quite compat-ible with the Darwinian view that the evolution of S2 is a gradual process of natural selection operating on genetic variation among the members of S1. (Devitt 2008: 372)

One might think that in the evolutionary change from S1 to S2, with distinct essences – G1 and G2, there is in effect a jump from one natural kind to another – even if the actual change within the lineage was gradual. But Devitt denies that there is any sharp line between the essences G1 and G2.

On the Essentialist picture, the evolution of S2 from S1 will involve a gradual process of moving from organisms that determinately have G1 to organisms that determinately have G2 via a whole lot of organisms that do not determinately have either. There is no fact of the matter about where precisely the line should be drawn between what constitutes G1 and what constitutes G2, hence no fact of the matter about where pre-cisely to draw the line between being a member of S1 and being a member of S2. Essences are a bit indeterminate. (Devitt 2008: 373)

This seems to suggest that there might be a time when the organisms within the lineage that was evolving were not part of a species at all, because they did not determinedly have either essence G1 and G2. Or perhaps these organisms had a little bit of each essence, and hence were a little bit one species and a little bit another. Depending on how fine the distinctions are between essences, this could be a very common state of affairs.

This response by Devitt demands an answer to the question how the essences G1 and G2 are determined – if they are distinct from the group of organisms in the way suggested. If groups of organisms vary gradually, and we determine essences from observation of these organisms, as he suggests, then we have to decide which time is determinative. In gradual change, there will be a difference in properties from one time to another. If we don't treat some particular time slice as determinative, then the essences will change as the individual members of the species change. Alternatively, is there some way to determine essential properties that is not based on a group of organisms at a particular time? Then we could have essences that are timeless as is typically assumed. But this seems to contradict the idea we get from Kripke and Putnam that science is in the business of determining the essences of natural kinds by observation of instances of the natural kind. The dilemma is this: Either we get our essences from the scientific investigation of organisms, and the essences therefore change as the group of organisms change, or we get our essences from some other non-empirical source, and they consequently have little to do with what science tells us about the organisms in question.

It may be possible to work out a way of thinking about these natural kinds with unchanging essences that also allows for gradual change and speciation within a lineage. That solution is sure to be labored though. The problem is that these essentialists are asking us to think about things that have a history and change over time, in terms of a metaphysics that does not obviously reflect change. If our metaphysics tells us that the basic, fundamental things are unchanging, but our science tells us that there is change, and that change is a fundamental part of nature, there is a discordance that counts against *either* the metaphysics or the science. Few biologists, or philosophers of biology for that matter, seem willing to sacrifice the science for the metaphysics. Is there a better way to think about the basic, fundamental nature of species that is amenable to evolution – that is historical and can accommodate change? Two approaches are more promising. One asks us to think of species as historical concrete, *individuals* that change over time. Another asks us to think of species taxa as *historical natural kinds*.

SPECIES AS INDIVIDUALS

Essentialism, in both its traditional conjunctive form, and its disjunctive cluster form, treats species taxa as *sets* (or classes) of things. On both approaches, the species sets are timeless in that the essential properties that determine set membership do not change over time. Evolutionary change then, must be from one natural kind set and its associated essence, to another natural kind set with its essence, rather than a change in the set and associated essence itself. To many traditional philosophers this seems plausible. Natural kinds like *water* and *electron* can be thought of as sets of things with essential properties. Change from one kind to another is common. Water can become hydrogen and oxygen, and vice versa. But to biologists and systematists this conception of species is much less plausible. The history of thinking about species laid out in chapters 2 through 4 reveals why. From Aristotle and Linnaeus, to Darwin and today, species have been conceived historically as lineages. A species taxon is a lineage of organisms connected by genealogy. This genealogical criterion is behind the theoretical concepts of Mayden and de Queiroz, which conceive species as general population lineages. How then can we think of species taxa at a more basic level and with respect to this historical component?

Michael Ghiselin began thinking about this question while serving a postdoctoral fellowship under Ernst Mayr in the 1960s. In 1965, he wrote a paper arguing that we should think of species as "individuals" (Ghiselin 1997: 14). He then developed that idea further in a 1969 book *The Triumph of the Darwinian Method*. Here Ghiselin first criticized essentialism for its apparent incompatibility with evolutionary change, and then criticized radical nominalism for its incompatibility with the reality of species. He proposed a third option, a "moderate nominalism" whereby species are *individuals*:

> It is possible to accept species as real and still embrace a kind of nominalism, if one looks upon species as individuals. Buffon (1707–1788), for example, would seem to have entertained the notion that a species is a group of interbreeding organisms. This point of view has certain analogies with the biological species definition of the modern biologist: "Species are groups of actually or potentially interbreeding natural populations, which are reproductively isolated from other such groups." A species is thus a particular, or an "individual" – not a biological individual, but a social one. (Ghiselin 1969: 53)

Ghiselin, no doubt influenced by Mayr's biological species concept that treated species as interbreeding groups, recognized that species (sexual species at least) were not just groups of organisms, they were groups with some sort of cohesion. Species that interbreed not only share genetic material, they recognize and interact with each other. For Ghiselin that meant they were true *populations*.

> The new manner of thinking about groups of organisms entailed the concept of a population as an integrated system, existing at a level above that of the biological individual. A population may be defined as a group of things which interact with one another. (Ghiselin 1969: 54)

The important idea here is that there is a level of biological organization above the level of the individual organism. This is obvious to us when we think about our own species – *Homo sapiens*. It is not just that individual humans share certain features – bipedalism, opposable thumbs, language and so on – they recognize each other as the same sort of beings, they look upon each other as potential mates and sometimes interbreed. Humans exchange information, gather in communal activities, dance and sing together, and do all of this largely *to the exclusion of members of all other species.* It is not just that we as humans are similar, we share our lives with other humans in all sorts of ways. Consequently, the species taxon *Homo sapiens* has a cohesion through a variety of social processes. This "process" cohesion is true of other species as well. Members of bird species sexually reproduce; they identify each other visually and through song; they cooperate to raise young; they fly together in flocks and so on.

Ghiselin developed this idea of species individuality in more detail in his 1997 book, *Metaphysics and the Origin of Species.* Here he argues that there are just two basic, fundamental kinds of things – individuals and classes (or "sets" in the terminology used here) (Ghiselin 1997: 37). Individuality is found in all sorts of things and at all levels, biological and otherwise.

> In the usual biological sense, "individual" is a synonym for "organism" but the ontological term is a much, much broader one. Although all organisms are individuals in the ontological sense, not all individuals in the ontological sense are individuals in the usual biological sense. We have suggested all sorts of things lacking the defining properties of "organism" that might be given as examples of an ontological individual. A chair is a piece of matter that an organism might sit on, and the world is full of such things. Or consider a part of an individual: one of a person's legs, or one of the legs of a chair. A part of an organism can be individual, including not just each and every organ, but each and every cell, each and every molecule, and each and every atom ... Likewise we can say that larger things can be individuals. An individual society would be a good example. If you do not like a society as an example of an individual, try the Earth, the Solar System, the Milky Way and the Universe. (Ghiselin 1997: 37–38)

As suggested in this passage, the idea of individuality extends both above and below the level of organism, to inanimate objects, as well as into human social domains. Individual soldiers and regiments in an army are individuals, as is the army itself.

One difference between classes (sets) and individuals is that classes have members or instances, while individuals have parts. Organisms are *parts* of species-as-individuals, whereas organisms are *instances* of species-as-sets (Ghiselin 1997: 38). Another difference is that individuals, unlike classes, are spatio-temporally restricted and continuous.

> [A]n individual occupies a definite position in space and time. It has a beginning and an end. Once it ceases to exist it is gone forever. In a biological context this means that an organism never comes back into

existence once it is dead, and a species never comes back into existence once it has become extinct. And although it might move from one place to another, there has to be a continuity across space as well as through time. (Ghiselin 1997: 41)

Classes (sets), unlike individuals, are spatio-temporally unrestricted and need not be continuous. While each individual chair in the class of chairs is spatio-temporally restricted with temporal continuity, the class itself is not restricted. Something can be a chair at any time and in any place. And there need not be a temporal continuity with others chairs. All chairs could be destroyed and then new chairs could be created at any arbitrary time afterwards (Ghiselin 1997: 42). By contrast, since individuals are spatio-temporally located and restricted, they have beginnings and endings, continuity and change over time. This is most obvious for individual organisms that have a birth, continuous existence, followed by death. Similarly, cells have a birth, continuity over time and a death in "apoptosis." Species taxa also have a birth, continuity and death in extinction or new speciation. *Homo sapiens* had a beginning, has continuity with change over time, and will undoubtedly have an ending.

The third distinction, according to Ghiselin, is that individuals are concrete, not abstract. In part this means that an *individual* can do things, and have things done to it. Individuals can participate in processes. An individual chair is a concrete thing and can support a seated person. It can be touched, lifted, moved, painted and broken. But no such actions can be performed to or by the abstract class of chairs (Ghiselin 1997: 43). Individual *organisms* can engage in processes such as reproduction or social grooming. Individual *species taxa*, like individual organisms, can participate in processes, in particular evolution and co-evolution. But an abstract class of organisms cannot do anything, and cannot engage in processes.

A fourth distinction is that individuals are not subjects *of* scientific laws – even if they are subject *to* laws.

[T]here are no laws for individuals as such, only for classes of individuals. Laws of nature are spatio-temporally unrestricted, and refer only to classes of individuals. Thus, although there are laws about celestial bodies in general, there is no law of nature for Mars or the Milky Way. Of course laws of nature apply to such individuals; they are true, and true of physical necessity, of every individual to which they apply. (Ghiselin 1997: 45)

There are no laws of *Homo sapiens*, for instance, although individual organisms of this species are still subject to all sorts of laws, beginning with the laws of physics and chemistry.

David Hull was initially critical of Ghiselin's *species-as-individuals* thesis (*s-a-i*), but came to accept it, partly because it made sense of the apparent lack of laws regarding biological taxa, such as *Panthera tigris* or *Homo sapiens* – just as there cannot be a law of a particular electron, a particular body of water or a particular planet (Ghiselin 1989: 53). Hull eventually took up Ghiselin's cause, arguing for, and extending, this thesis. Hull agrees with Ghiselin on the basic idea of the individual and its contrast with classes:

> By "individuals" I mean spatiotemporally localized cohesive and continuous entities (historical entities). By "classes" I intend spatiotemporally unrestricted classes, the sorts of things which can function in traditionally defined laws of nature. The contrast is between Mars and planets, the Weald and geological strata, between Gargantua and organisms. (Hull 1992b: 294)

Like Ghiselin, Hull extends individuality beyond species and biology, to social groups, theories, concepts and more.

EVALUATING INDIVIDUALITY

Philosophers have been somewhat less inclined than systematists and biologists to adopt the species-as-individuals thesis. One standard philosophical objection has been motivated by the disanalogy between individual organisms and species taxa with respect to cohesion and causal integration. Individual organisms have cellular structure, and typically have extensive causal integration. In vertebrates, lungs (or gills) and hearts interact as part of a cardiovascular system; muscular systems are supported by skeletal structures; and gastrointestinal systems provide chemical nutrients to all the other systems, and eliminate waste products. Vertebrates require this causal integration to live. Disrupt these systems, and death is the likely result. Species, on the other hand, seem to have much less causal integration and as a result can survive the loss of many of their members and the relationships among members. Michael Ruse focuses on this disanalogy.

> We think organisms are individuals because the parts are all joined together. Charles Darwin's head was joined to Charles Darwin's trunk. But in the case of species, this is not so. Charles Darwin was never linked up to Thomas Henry Huxley. Of course, you might object that although Darwin's head was never linked directly to his feet, they were linked

indirectly through intermediate parts. Analogously, as evolutionists presumably we believe that Darwin and Huxley were linked by actual physical entities (namely the succession of humans back to their shared ancestors). But, this objection fails, for the point is that these links have now been broken and lost. If (gruesome thought!) Darwin's head were physically severed from his feet, we would certainly have no biological individual. (Ruse 1987: 232)

Surely this commonsense objection is decisive. Species, whatever they are, lack the cohesion and causal integration of individual organisms. If so, there is a problem with the species-as-individuals thesis.

While philosophers have typically found this *cohesion argument* compelling, biologists are generally less convinced. Mayden and de Queiroz, for instance, both endorse the species-as-individuals thesis. According to Mayden, the theoretical significance of the Evolutionary Species Concept is due in large part to the fact that it treats species as individuals rather than classes (Mayden 1997: 415). De Queiroz similarly emphasizes the similarity between organisms and species:

Organisms and species are not only individuals; they are very similar kinds of individuals in that both are lineage segments ... Indeed, one could even go so far as to say that organisms and species (along with genes and cells) are members of the same category of individuals – lineage forming biological entities – although they obviously differ with respect to the level of organization. (de Queiroz 1999: 67)

Those who work in the biological sciences typically don't see such a distinctive and important disanalogy between individual organisms and individual species taxa. There are, I believe, several reasons based on familiarity with biodiversity, tradition and disciplinary practices.

When philosophers think about individual organisms, they have tended to focus on large vertebrates. Typically, they are *most* interested in what constitutes *human* identity and individuality. Biologists, in contrast, tend to consider the full range of organisms from bacteria and fungi to vertebrates. Hull notes this difference between philosophical and biological stances, and endorses the biological:

Differences between these two analyses have three sources: first, philosophers have been most interested in individuating persons, the hardest case of all, while biologists have been content to individuate organisms; second, when philosophers have discussed the individuation of organisms, they have usually limited themselves to adult mammals, while biologists have attempted to develop a notion of organism adequate to handle the

163

wide variety of organisms which exist in nature; and finally, philosophers have felt free to resort to hypothetical science fiction examples to test their conceptions, while biologists rely on actual cases. In each instance, I prefer the biologists' strategy. A clear notion of an individual organism seems an absolute prerequisite for any adequate notion of a person, and this notion should be applicable to all organisms, not just a miniscule fraction. (Hull 1992b: 301)

Part of the problem is that the philosophical perspective is misleading:

Given our relative size, period of duration, and perceptual acuity, organisms appear to be historical entities, species appear to be classes of some sort, and genes cannot be seen at all. However, after acquainting oneself with the various entities which biologists count as organisms and the roles which organisms and species play in the evolutionary process, one realizes exactly how problematic our commonsense notions are. (Hull 1992b: 295)

A full consideration of biodiversity reveals the bias in philosophers' commonsense notions of individuals, and its focus on vertebtrates and humans. If Hull is right, the disanalogy philosophers tend to find between organisms and species is based on an incomplete consideration of the full diversity of life.

When we follow Hull's suggestion and look at the full range of biodiversity, the cohesion argument loses some of its force. Individual organisms are not all as cohesive as vertebrates. Ghiselin explains:

A situation in which an organism breaks up into component parts that never get back together again is familiar even to the lay person. Propagation by cuttings, budding, and fission of some animals such as starfish are good examples. There are fewer examples of organisms that break up, then fuse back together, but slime-molds are an example. These fungi, which form lineages produced by asexual reproduction, forage independently on organic materials. Later in their life cycle they come together to form a single mass, complete with reproductive organs that give rise to spores. Sometimes they are called "social amoebae," and the term aptly compares them to societies that are united only from time to time. (Ghiselin 1997: 55)

If we look throughout biodiversity, beyond the more integrated vertebrates, we see a wide range of integration and cohesion, from what seem to be social organisms, made up of colonies of cells, such as we see in slime-molds and fungi, to moderately integrated organisms that still have social elements, such as we see in jelly fish, to organisms that

164

can bud or divide, all the way up to the highly integrated vertebrates. Given a different biological perspective, we might treat these less cohesive organisms as paradigmatic individuals. If so, then the relative lack of cohesion within species would seem less problematic. Perhaps we should follow Hull's advice in the passage quoted above, when he recommends that we first get a good understanding of the term *individual* as it applies *generally throughout biodiversity*, and then we can apply it to unique organisms such as humans.

COHESIVE CAPACITY

The second reason the cohesion argument has been less compelling to biologists than philosophers is that biologists are typically more cognizant of the variety of cohesion processes in species. We can think of cohesion processes negatively, in terms of the processes that isolate the organisms of one species from another, or positively, in terms of those processes that cause cohesion among the organisms of a species. Often a cohesion process does both. Interbreeding and gene flow are the most obvious and perhaps most significant cohesive forces. It is both isolating, in that it does not typically occur across species, and cohesive, in that it occurs within species. This particular *cohesive capacity* is more complex than usually recognized. Often it is assumed that there is a clear distinction between sexual and asexual organisms, and that only sexual organisms have this cohesive capacity. But some organisms may reproduce sexually under some environmental conditions and asexually under others. Some green algae, for instance, reproduce asexually when hydrated, and sexually in dry phases (Niklas 1997: 42). Here there is a *capacity* for cohesion by gene flow but not always an actual gene flow. Gene flow can also occur over great distances, especially among marine mammals (Ghiselin 1997: 102). And even in the plant kingdom, there can be sexual reproduction and gene flow over distance (Niklas 1997: 27).

The factors that produce gene flow within a species, and isolate the individuals of a species from other species, are cohesive in both the positive and negative senses. The mate recognition systems of individual organisms, for instance, key in on just those individual organisms that are potential mates, and distinguish them from those that are not, based on chemical communication and behavior. Territorial and courtship displays among birds are typically species specific, as are some calls among primate species. The head-bobbing movements of spiny lizards are species specific and allow the females to identify the males of their own species and avoid the males of other species (Wilson 2000: 183–184). In each of these cases these species-specific ways of communicating give cohesion to the species, but also isolate the members of the species by distinguishing and excluding individuals of other species. And while many of these cohesive factors are related to sexual reproduction, many are not. Conspecifics recognize, communicate and interact with each other in ways they don't with individual organisms of other species, resulting in social spacing effects, social symbiosis, dominance and caste systems and more. There are, for example, a variety of density-dependent effects, whereby the density of conspecifics affects emigration rates, and developmental trajectories. At higher densities, caterpillars of the cotton leaf worm *Spodoptera littoralis* become darker, more active and produce smaller adults. In many aphid species, adults develop wings and turn from parthenogenesis to sexual reproduction, and disperse (Wilson 2000: 83). But the most striking effects are perhaps found in the "plague" locusts:

> When these insects are crowded during periods of peak population growth, they undergo a phase change that takes three generations, from the *solitaria* phase that is first crowded through the intermediate *transiens* in the second generation to the *gregaria* phase in the third generation. The final adult products are darker in color, more slender, have longer wings, possess more body fat and less water, and are more active. In short, they are superior flying machines. Also, their chromosomes develop more chiasmata during meiosis, resulting in a higher recombination rate and, presumably, greater genetic adaptability. Finally, both the nymphs and the adults are strongly gregarious, readily banding together until they create the immense plague swarms. (Wilson 2000: 83)

These cohesive effects are just some of the many we see in nature from the social insects, to fish, frogs, birds, prairie dogs and dolphins, to elephants, dogs, lions and primates.

For obvious reasons, this social cohesion is most apparent to us in our own species. Not only do we identify other humans as potential mates, we also identify them as individuals with whom we can potentially communicate through language – even if we speak different *human* languages. We share culture and have cultural exchange with other humans, but not with individuals of other species. We engage in a variety of activities predominantly with other humans. We play games, dance and eat with other humans in ways we don't with members of other species. If we

think about the ways that humans have social cohesion – the many processes that govern our interaction with other humans, and rule out similar
interaction with members of other species – perhaps we can get a better
sense of the full range of ways that other species can have cohesion.

Selection processes can also have cohesive capacities within species.
Most obvious is sexual selection. Among those species that reproduce
sexually, there are typically tendencies and preferences that govern mating. The preferences of the females who choose males, for instance, may
affect the mating behaviors and the selection forces on males. The males
that compete for females are likewise subject to selection on the basis of
interaction with male conspecifics. Natural selection in a common environment can maintain homogeneity and stasis within a population. Just
as predation can change the color distribution among moths, it can prevent it from changing. This homeostasis is then a source of species cohesion (Hull 1992b: 300). Co-evolution can similarly affect the cohesion of
a species. Members of one species may form co-evolutionary relationships with a species of bacteria or parasites.

It may be objected, and rightly so, that many of these forces of cohesion do not operate at just the species level. Organisms come grouped in
breeding pairs, families, colonies, demes, populations and metapopulations. It may be that there are different levels of cohesion at different levels of grouping. This objection is surely correct, and reinforces
the claims of Ghiselin and Hull that individuality occurs at many levels
of biological organization. If so, then the question is: what are the levels of cohesion that correspond to the species level? This is a complex,
empirical issue, and I cannot say anything conclusive here, other than
to note that the identification of species level cohesive effects is a significant part of what has driven the proliferation of operational species
concepts. We have ecological, interbreeding and mate-recognition concepts precisely *because* there are cohesive processes that bind members
of a species together. And when cohesion at the species level fails, we
get new species (Ghiselin 1997: 100)! Whatever the case, if we focus
on these cohesive processes, the disanalogy between individual organisms is weakened. Whether or not this is sufficient to refute the cohesion argument is debatable. Perhaps cohesion comes in degrees, as Hull
argues:

> Spatiotemporally organized entities can be arrayed along a continuum
> from the most highly organized to the most diffuse. Organisms tend to
> cluster near the well-organized end of the continuum while species tend
> to cluster near the less organized end, but as Mayr notes ... there are

entities commonly classed as organisms that are no better organized than
are many species. (Hull 1989: 114)

And if cohesion comes in degrees, then perhaps individuality also comes
in degrees. Nature would therefore have *individuality* at many levels and
of many degrees.

SPECIES AS HISTORICAL KINDS

As we noted at the beginning of the last section, philosophers are less
inclined than biologists to accept the species-as-individuals thesis.
Relatively few philosophers, however, have been willing to adopt the
traditional property essentialism that conceives species as natural kinds,
on the basis of the possession of a set of *intrinsic* essential properties.
This is particularly true of the conjunctive essentialism advocated by
Devitt and Kitts and Kitts. Nonetheless, many philosophers are still
inclined to see species taxa as natural kinds, but instead as *historical
natural kinds*, with historical essences.

Traditional property essentialism treats species essences as intrinsic properties, usually morphological or genetic. Intrinsic properties are
roughly those properties that don't require reference to anything external.
Size, shape, color and genetic composition are typically taken to be intrinsic. Extrinsic or relational properties, on the other hand require reference
to something external. Being a parent or sister are extrinsic or relational
properties because they are dependent on something external – a child or a
sibling. There are, to be sure, doubts that this distinction between intrinsic
and extrinsic properties is clear and unambiguous. We might for instance,
think coloration to be relational property rather than intrinsic, because
the color a thing has, is dependent on the light of the environment and the
perceptual apparatus of a perceiver. Fish in deep water for instance, have
little or no color, until illuminated by an artificial light. And for humans,
with a particular range of sensitivity to light wavelength, the color seen
will likely be different than for other organisms with different sensitivities.
Color is therefore relational to subjects and photic environments (Richards
2005: 273–276). Nonetheless, and in spite of these complications, there is
still a significant and relevant distinction between intrinsic and relational
properties that serves as the foundation of historical kinds.

The idea of species as historical kinds is based on the same idea that
makes the species-as-individuals thesis plausible. As we saw in previous
chapters, naturalists have long been inclined to use genealogy in species

groupings. An organism is a member of a species if its parents were members of that species. We saw this same criterion for species membership in the views of Buffon, and Darwin's contemporaries, as Darwin noted in a passage quoted in chapter 3: "With species in a state of nature, every naturalist has in fact brought descent into his classification" (Darwin 1964: 424). This practice of grouping on the basis of descent might plausibly be regarded an essentialist approach, but employing relational rather than intrinsic properties as essences. An individual is a member of a species based on parentage. Charles Darwin was a member of *Homo sapiens* because he was born of members of *Homo sapiens*. And if he were not born of human parents, he would not be a member of *Homo sapiens*. Being an offspring of humans is then an essential, relational property.

We find this sort of relational and historical essentialism in the views of Michael Ruse, who asks why his dog Spencer is of the species *Canis familiaris*: "So why do we want to say that he is part of the species? Because he descended from the original ancestors, along with the rest of the group – that's why ... Descent is starting to look very much like an essential property" (Ruse 1987: 236). Paul Griffiths and Joseph LaPorte have followed Ruse's lead here, arguing for an historical essentialism. Griffiths begins with the idea of homeostatic property clusters, but doesn't see it as restricted to intrinsic properties (Griffiths 1999: 218). For Griffiths, this more general approach allows for essences that are historical and relational:

> Nothing that does not share the historical origin of the kind can be a member of the kind. Although Lilith might not have been a domestic cat, as a domestic cat she is necessarily a member of the genealogical nexus between the speciation event in which the taxon originated and the speciation or extinction event in which it will cease to exist. It is not possible to be a domestic cat without being in that genealogical nexus. (Griffiths 1999: 219)

The *essence* of a species, then, is its location within the evolutionary tree. The members of a species must have the relevant relations to evolutionary ancestors, and it is this evolutionary history that allows us to explain, and thereby understand, the important features associated with a species or higher taxon – morphology, genotype, behavior and so on.

Joseph LaPorte similarly endorses a conception of species as historical kinds with historical essences, but instead starts with the idea behind the species-as-individuals approach:

> If species must be historical, as individuality theorists and a growing majority of biological systematists and philosophers of biology say, then

the view, popular in the philosophical community, that each species is
identical to a genetic kind is mistaken. An Alpha Centaurian animal genetically identical to the horse would belong to a different species. (LaPorte
2004: 11)

But that there are no property essences in terms of genetic traits does not
imply that there are no essences. There are historical essences: "Given
that biological kinds are delimited historically, the essences of kinds
simply become historical" (Laporte 2004: 64).

Notably, however, LaPorte does not see this as ruling out the species-
as-individuals theses. These are two different, yet compatible, ways of
conceiving species.

> Let it be *granted* to s-a-i theorists that the organisms of a species con-
> stitute an individual, either because they display cohesion after all or
> because individuality, or individuality of the relevant stripe, does not
> require cohesion ... Even if it is granted in this way that there is an indi-
> vidual whose parts are organisms of a species, it is nevertheless the case
> that there is a kind here as well. (LaPorte 2004: 17)

This passage may seem puzzling, in that at some level LaPorte sees no
conflict between these two metaphysical stances. We can talk about species taxa as historical kinds – or we can talk about them as individuals.
But if we can talk about species in either way, why is there even a debate
about the metaphysical status of species?

SETS VERSUS INDIVIDUALS

The main question of this chapter is about two fundamentally different
ways of thinking about biological species. Should we conceive species
as sets (or classes) of things defined by a set of properties, or as spatio-
temporally restricted individuals? Or should we follow the suggestion
of LaPorte and treat species as *both* sets and individuals? To answer
these questions we need to address two others. First, what is at stake –
what is the significance of the metaphysical question about the nature of
species? Second, what are the relevant evaluative criteria – how do we
decide which general conception is to be preferred? At the beginning of
this chapter, I contrasted two ways to approach metaphysics: transcendent and naturalistic. If we adopt the naturalist approach, which sees
metaphysical questions as internal to science, and continuous with the
empirical and theoretical considerations of the relevant science, we can

have something positive to say about the metaphysical conceptions of species, by reference to science.

Laporte is surely right in his observation that we *can* talk about species taxa using either *sets* vocabulary or *individuals* vocabulary. But we *can* also talk about species as natural kinds with intrinsic property essences – as contorted as that may be in light of evolutionary change. Furthermore, we *can* talk about species with clusters of properties as essence, and we *can* talk about them as historical kinds with relational properties. But we *can also* talk about species using a creationist and theological vocabulary – as if they were products of Divine design. Analogously, we can talk about organisms as sets of cells. (See Sober 1984: 337.) We could, for instance, define the individual with the name Plato as the set of cells that had a historical origin in the fertilization of the egg that produced this individual. Or we could identify the set of cells that comprised the individual Plato in terms of clusters of properties representing the homeostatic mechanisms that result in the development and continuation of Plato the individual organism. This might be a difficult project to carry out in its details, but surely it is possible in the same way it is possible to treat species as historical or cluster kinds.

The reason we can talk about organisms and species in either way – as sets or individuals – is that the dispute is not directly resolvable through observation or experiment. We cannot just look at species taxa and see whether they are sets or individuals, just as we cannot look at organisms and see whether they are sets or individuals. This doesn't mean that there aren't good reasons to regard species and organisms one way or the other. If we accept the premise that metaphysics is continuous with science, then we can ask the following question: how *should* we conceive species given what we know about the species and their evolution? David Hull argues for a coherence criterion:

> Empirical evidence is usually too malleable to be very decisive in conceptual revolutions. The observation of stellar parallax, the evolution of new species before our eyes, the red shift, etc. are the sorts of things which are pointed to as empirical reasons for accepting new scientific theories. However, all reasonable people had accepted the relevant theories in the absence of such observations. Initial acceptance of fundamentally new ideas leans more heavily on the increased coherence which the view brings to our general world picture. If the conceptual shift from species being classes to species being historical entities is to be successful, it must eliminate longstanding anomalies both within and about biology. (Hull 1992b: 306–307)

A metaphysical position can cohere more narrowly with evolutionary theory, or more broadly relative to an overarching metaphysics. In the final chapter, we will briefly look at the broader question. For now we will look at the narrower question: does the species-as-individuals view cohere better with the evolutionary framework than the species-as-sets view?

Hull does not here say what it means for a conceptual framework to cohere with either a scientific theory or a world picture, but there are plausible ways we can think about coherence. First, and most obviously, coherence seems to require consistency. Is either the individuals or the sets approach straightforwardly inconsistent with evolutionary theory? As we have seen, the traditional conception of species as natural kinds can be made consistent with evolutionary change, but only in a highly problematic way. As Devitt explains it, a lineage of organisms passes from one unchanging natural kind through an indeterminate area to another natural kind with a different and unchanging essence. At some point, a lineage would seemingly be between essences and natural kinds – unless the essences changed with it. Perhaps we can avoid this inconsistency of evolutionary change with unchanging essences, if we adopt the cluster set approach. Boyd allows that the cluster of essential properties can change over time, although paradoxically this does not change the "definition" of the species (Boyd 1999: 144). Finally, the conception of species as historical kinds connected by genealogical relations is obviously consistent with evolutionary change, as it says nothing about the morphological and genetic properties of particular organisms.

The idea that species are individuals is straightforwardly consistent with evolutionary change. One of the things individual organisms do is undergo a process of development and change form birth to death. Similarly, one of the things segments of population lineages do is undergo a process of change from birth to death. But there are worries about consistency with evolutionary processes, natural selection in particular. Because natural selection operates *within* a species taxon (as well as among species), there is potentially a fundamental evolutionary process that works counter to individuality. Ruse raises this worry as part of his cohesion objection:

> [T]here is something very odd indeed about speaking of a species as an individual. It is very far from being an integrated unit like an organism. The individual organisms of a species are all working for their own benefits, against those of others. Any species cooperation, any species integration,

is secondary on the particular organisms self interest. And, in any case, one is hardly likely to get species-wide secondary effects. Cooperation will at most, be between relatives, or fellow population members. Generally, selection pits organisms against each other (although not necessarily in a crude "nature red in tooth and claw" fashion) ... Individual selection and the s-a-i thesis simply do not go together. (Ruse 1992: 351)

Ruse is surely right that the focus of natural selection is at the level of the individual organism and that this seems to promote individuality at a level lower than the species. But this must be counterbalanced against the wide variety of factors that are cohesive: sexual selection, the stabilizing effects of natural selection, interbreeding and gene transfer, and the various sources of social cohesion described earlier in this chapter. Moreover, there are some who think that species selection, based on species level traits, may operate, and sometimes contrary to individual selection. Whatever the case, the species-as-individuals thesis does not imply that there is no cohesion at other levels, even if that cohesion reduces species level cohesion. It may even be that natural selection has complex and sometimes conflicting implications for cohesion at different levels of biological individuality.

There is no conclusive answer yet about which metaphysical position is more coherent with evolutionary theory. Perhaps this will be answered in the actual practice of scientists over time. I suspect, but cannot prove, that there is a developing consensus in favor of the species-as-individuals thesis. Many of those who work in the relevant biological disciplines are becoming more inclined to the individuality conception, including Mayden, and de Queiroz. In part this might be based on the *fertility* of the respective approaches. We can ask of both the individuals and sets approach: what fruitful questions does it raise, what new ideas does it suggest, what new experiments does it propose? In general, we can ask how a particular metaphysical stance helps us to think about the theoretical framework and empirical phenomena. To see the advantages of the individuals approach on this heuristic criterion, we can return to the views of the most prominent advocate of the species-as-sets conception, Philip Kitcher.

As we saw in chapter 5, Kitcher argued for an approach I identified as a pragmatic pluralism. According to Kitcher:

Species are sets of organisms related to one another by complicated, biological interesting relations. There are many such relations which could be used to delimit species taxa. However, there is no unique relation which is privileged in that the species taxa it generates will answer to the needs of all biologists and will be applicable to all groups of organisms. (Kitcher 1992: 31)

My emphasis in that chapter was on its pluralism. Here we can focus on its assertion that species are *sets* of organisms. There are two main criteria for forming these sets, according to Kitcher: structural and historical. Which way of dividing and grouping organisms into sets is relevant depends on what questions are asked and what sorts of explanations are appropriate. There is little doubt that we could divide and group organisms into sets on the basis of theoretical interests and "interesting biological relations," but *which* of these interesting biological relations will generate *species taxa*? Many of them will not. Elliott Sober criticizes Kitcher's proposal on these grounds:

> It isn't that evolutionary biology has no interest in identifying natural kinds. We already have a number of nice candidates. Predator may be one; sexuality may be another. But these are not species concepts. It is no surprise that these natural kinds cross-classify each other; presumably there is an overlap relationship between the predator populations and the populations that reproduce sexually. (Sober 1984: 335)

The problem here is that this framework does not give us guidance in how to think about species taxa specifically. There are many properties that are biologically interesting, and worthy of investigation, but do not group organisms into *species* sets. We can for instance look at sets of organisms with particular genes, or particular behaviors, or particular social structures. These are all potentially of interest, but they are not necessarily relevant to species groupings.

This is also a problem for cluster sets and historical sets. It may be that there are homeostatic mechanisms that are potentially relevant for each species taxon, but there is no set of criteria that will be relevant to all species. Sexual reproduction is relevant to some, not to others. Furthermore, there may be mechanisms that are relevant to all organisms, but not to *species* sets. The constraints placed on growth from basic physics are relevant to plants and trees as well as vertebrates. A tall tree and a giraffe both have constraints based on gravity. But these factors are not necessarily relevant for species groupings. Gene regulation networks are often homeostatic mechanisms in some sense, but different species can share particular networks. Treating species as cluster sets is possible, but it gives us no guidance on what should count as a relevant *species* property, and how these properties should cluster.

Similarly, while it is true that genealogical relations are required to be a member of a species, not all genealogical relations are relevant. All creatures (as far as we know) are related via the tree of life. But not all creatures are of the same species. Speciation events occur that produce new species. And over speciation events, there are historical relations, but there are also new species. Historical sets do not give us guidance about deciding when a historical relation is necessary and sufficient for set inclusion and when it is not. And there may be no single way of making this distinction for all species. So while there is no doubt we *can* think of organisms as sets of things, with various properties determining set inclusion, the choice of criteria for *species* set membership is indeterminate. Like the cluster approach, this conception simply cannot give us guidance here.

The *species-as-individuals* approach, on the other hand, seems to give all sorts of guidance in thinking about species. First, it tells us that species are populations with various forces of integration and cohesion. That tells us to look at *whatever* might have some integrating function or other. In some species it is sexual reproduction and mate recognition systems. In others it might be environmental. The *species-as-individuals* approach also tells us that species have birth and death. We can therefore look at *whatever* processes may result in speciation – polyploidy, geographic isolation and so on. And if species "death" is due to new speciation, these same processes will be relevant. If sometime it is mere extinction, then the processes that operate in extinction will be relevant. If we accept the idea that species taxa are historical and have population structure, then thinking about them as individuals is a source of guidance for species grouping as well as further investigation. This is not just a heuristic value, although it is that, it is also of value for the further development of evolutionary theory, and the clarification of the species concept.

The insight from the division of conceptual labor solution to the species problem can perhaps help us here on the relation between the conceptions of species as individuals and sets. We could treat the *species-as-individuals* stance as the primary metaphysical stance, and the *species-as-sets* as a secondary stance that helps us think about processes that may – or may not – function in the integration of populations, or the segmentation of species through speciation events. The *species-as-individuals* stance gives us guidance in thinking about species taxa, in developing our theories, and pursuing new research. Along the way, there will be factors that are important in some species, or in various processes. These factors can be the basis of the sets of organisms that form natural kinds. The set of all sexual organisms, for example, does not pick out a species taxon, but can form other sets for various theoretical reasons. Here we can have sets based on all of the theoretical interests we may have, and representing all the interesting biological relations there may be, but these sets are not thereby determinative of species taxa. A metaphysical framework then, can have a hierarchical structure just as can the less general theoretical framework. The bottom line is that this metaphysical framework based on *individuality* promises to be a fertile way to think about species within the evolutionary context that coheres with what evolutionary theory tells us about species and that is confirmed by empirical investigation.

CONCLUSION

In this chapter we have looked at a metaphysical question: given that species taxa have various characteristics and function within evolutionary theory in specific ways, what is the best general, fundamental way to think about them – as sets or individuals? The species-as-individuals answer seems most promising in terms of coherence with evolutionary theory, fertility in thinking about species taxa, and in the development of evolutionary theory. If the species-as-individuals conception is so superior for its value to biological thinking, why is the *species-as-sets* conception so attractive to philosophers? There is, I believe, a clash of two disciplines here – philosophy and biology – with different goals, methods, tools and traditions. Natural kind and set thinking has been an important part of philosophy for a very long time – at least since Plato. Because philosophers have been trained in this tradition, it is natural for them to turn to it to make sense of the world. And most philosophers know how to play the definition game: propose a definition in terms of essential properties, test that definition against intuitions, and then revise the list of essential properties to account for the counterexamples. As the saying goes, for the man with only a hammer, everything becomes a nail. To be fair, it is surely legitimate for philosophers to try to extend a particular philosophical approach as far as possible. This is what philosophers have done in trying to conceive biological species as natural kinds and sets.

Contemporary biologists and naturalists come from a very different tradition, one that is permeated with the realization that nature is messy, complicated and difficult to characterize in any systematic way.

Linnaeus, as we saw in chapter 3, was acutely aware that his system oversimplified nature, and was therefore unnatural. The same goes for Buffon, who was even more impressed with the complexity of nature, and the difficulty of generating formal representations of nature. But even more important than this awareness of biological complexity, biologists are deeply steeped in evolutionary theory. This theory is central to biology, as implied by the oft-quoted claim of Dobzhansky that "nothing in biology makes sense except in light of evolution." For biologists, coherence with evolutionary theory is a primary consideration. A metaphysical conception employed in biological practice must work well *within* the evolutionary framework. My bet is on the historical, *species-as-individuals* metaphysical framework.

7

Meaning, reference and conceptual change

I have so far been engaged in two projects. The first project, carried out in the first four chapters, is descriptive and aimed at understanding the many different ways of thinking about species from Aristotle to modern systematics. The second project, carried out in chapters 5 and 6, is primarily prescriptive, aimed at understanding how we *should* think about species – given what evolutionary theory tells us about them. One insight of the prescriptive project, based on the idea of the division of conceptual labor, is that to understand and evaluate a species concept we need to look at a level above the concept, into how that concept functions within a particular conceptual framework. But we can also look at the nature and functioning of an individual species concept. In part, this means understanding how the term *species* has meaning and refers to things in the world. To do so, we will first look at some standard philosophical views about concepts – what they are and how they work. What we shall find is a complex feedback process between meaning and reference, a definitional structure and reference potential determined by theory, a set of social factors relating to the social structure of science, and a set of practical values governing the use of the term *species*. We will first begin with a brief sketch of some prominent philosophical views about concepts. We will then return to the descriptive project, to understand how the Greek *eidos* and the Latin *species* have been used from Aristotle to modern systematics. This will give us the resources to draw some general conclusions about the nature and functioning of species concepts and how that has contributed to the species problem.

Up to this point, I have assumed that the idea of a scientific concept is clear enough. I have also assumed that it is possible to analyze the species problem and give a solution in the terms used by those who have engaged in the debate about species. This may be so if we are just interested in the ways biologists have defined and applied the various species concepts. But a deeper philosophical understanding into meaning, reference and conceptual change requires a closer look at the nature of concepts and how they work. For instance, if concepts are mental representations that have the character of images, as John Locke, David Hume and many of the early modern philosophers thought, then meaning can be understood in terms of the nature of the mental image. Conceptual change would then be a change in the representational image. Few philosophers think this image version of the representational theory of concepts to be adequate, although many think some representational theory or other is correct (Margolis and Laurence 2006: 3). But our interest here is not in what biologists are representing to themselves (if they are in fact representing something to themselves) when they think of the species concept or species taxa, but how these concepts function in science. For purposes here we must look to other ways of thinking about concepts, conceptual meaning and conceptual change.

A second way of thinking about concepts is based on the ability to make the relevant discriminations. Having the concept of *cat*, for instance, means having the ability to discriminate cats from other things (Margolis and Laurence 2006: 4). By extension, having the concept of *species* means having the ability to distinguish species things from other things. This is certainly relevant to our discussion of the species problem, but it is only part of the picture. We are also concerned with the meaning of the term *species* and the meaning of various species terms such as *Homo sapiens*. For this we can turn to the third way of thinking about concepts, in terms of a "Fregean sense."

In a classic German paper of 1892, translated into English as "On Sense and Nominatum," Gottlob Frege addressed the question of how language can represent things in the world. Two principles advocated there have become central to modern discussions of concepts. The first principle is that linguistic entities such as concept terms function in propositions in two ways: first, through a "nominatum," what the term *refers* to, what it *designates* or *denotes*; second, through the "sense," or *meaning* of the term. According to Frege, the sense of a term is grasped by anyone who knows the language, and is to be identified with the *description* that would be associated with the term in that language. The sense or meaning of the term *water*, for instance, would be identified with an associated description of *water*. The meaning of a term must be distinguished from what it refers to, or denotes, because, according to Frege, co-referential terms (terms that refer to the same thing) often have different meanings. Two terms that referred to the planet Venus, for instance, have different meanings based on the descriptions that designate different times of appearance in the sky: "The nominata of 'evening star' and 'morning star' are the same, but not their senses" (Frege 1990: 191). Meaning is therefore more fine-grained than reference, in that two terms can refer to the same thing, yet still have different meanings.

The second principle Frege advocated is that the meaning (sense) of a term determines its reference. The description associated with a term's meaning picks out what the term refers to or denotes (Frege 1990: 191). So the description associated with the *morning star* as the star that appears in the morning just above the horizon and in a particular place and with a particular color and so on determines the reference of that term – the planet Venus. And the description associated with *water* determines what that term refers to or denotes. *Water* just refers to the stuff that has the descriptive properties associated with the term *water*. Sense or meaning, then is associated with a description which then determines reference. Because of this priority of description over reference, the Fregean approach is often identified as a type of "descriptivism" (Soames 2005: 7).

DEFINITIONAL STRUCTURE

If meaning is to be understood in terms of an associated description, it is natural to ask about the nature of the description. One standard answer conceives the description in terms of a definition with a particular structure. According to this "classical theory," concepts have definitional structure in terms of a set of necessary and sufficient conditions for falling under the concept. The meaning of the concept term is then a definition consisting of a set of singly necessary and jointly sufficient conditions (Margolis and Laurence 2006: 8). As we have seen in previous chapters, this approach has become associated with the essentialist species concepts of the Essentialism Story.

There are, however, other ways to think about definitional structure. One is "probabilistic" in that a strict possession of a set of necessary and sufficient conditions is not required. Rather, something can fall under a concept to varying degrees depending on how many conditions are met, and how typical or characteristic the particular conditions satisfied are (Murphy and Medin 1985: 294). This way of thinking about concepts as probabilistic clusters of conditions has lead some to advocate a "prototype" or "exemplar" approach, where some instance of the concept that instantiates the core set of conditions comes to represent it as an exemplar or ideal instance (Murphy and Medin 1985: 295). This suggests there are degrees of concept application. Something can more or less fall under a particular concept depending on how many and which conditions are satisfied, or how close the analogy is with the exemplar. Definitional structure here might include a conceptual core that has greater definitional weight than other conditions, without thereby constituting a set of necessary and jointly sufficient conditions. The meaning of a term would then be some weighted cluster or other of the descriptive properties or conditions associated with the concept.

Because this way of thinking about structure does not rely on a sharp and fixed distinction between the necessary and sufficient conditions usually associated with definitions, and those conditions that are merely "accidental," it is not strictly speaking a view just about definitions, but more generally about the descriptions that constitute the Fregean meaning of the term. For this reason, the more general term *descriptive structure* might be more appropriate here than *definitional structure*. It should also be noted that on the classical structure, with its distinction between necessary and sufficient definitional conditions and accidental conditions, some parts of the descriptive structure may be relevant to meaning, but nonetheless not definitional. Consequently, it may be misleading here as well to speak of *definitional structure* as constitutive of meaning. I will however, continue to use this term with the caveat that it does not assume that all of the descriptive content is definitional, or even that there is a clear and fixed distinction between the definitional and non-definitional content.

There are two questions that these theories of definitional structure cannot obviously answer. The first question is about the formulation of the definition, or more generally, the description associated with the meaning of a concept. What guides the selection of definitional or descriptive conditions that is associated with the meaning of the term (Murphy and Medin 1985: 296)? The classical and probabilistic theories of structure only tell us *how* the conditions are structured, not *what* conditions to include – the content. The second question is about the relation among the conditions of the definition and description. Are the conditions included in the definition or the descriptive content of a concept related? If so, how and why? Or, in other words, is the definitional structure coherent, and, if so, what makes it coherent? Here as well, it is unclear how the classical and probabilistic theories of definitional structure can answer this question. Since they don't give guidance to the *content* of the definitional structure, it is not clear they can tell us about a feature of the content – the relationship between the various descriptive and definitional conditions. What is required is a theory of definitional or descriptive content.

THE "THEORY THEORY"

The last several decades has seen the development of an approach to concept meaning known as the "theory theory." The idea behind this approach is that the definitional structure of concepts is filled out and made coherent by some *theory*, scientific or otherwise, that contains the relevant concept. Chemical theory, for instance, gives the definitional conditions of the concept of *water*, based on its molecular composition of two hydrogen atoms and one oxygen. Other descriptive conditions provided include freezing point, density, appearance and so on. These conditions *cohere* because the concept of water is given its meaning – definitional and descriptive content – by a chemical theory that identifies what attributes or conditions are important and how they are related. 'Present on Earth but not the moon,' for instance, is an attribute of water that is unimportant, according to chemical theory and is therefore not included in the definitional content.

Many of the advocates of the *theory theory* are interested in applying this approach beyond our immediate interest here, in particular to the use and development of concepts in children. This empirical emphasis is not surprising since many of those who advocate the *theory theory* are psychologists. Nonetheless, its relevance to the history and philosophy of science is widely recognized. This is partly due to the fact that this approach is grounded on the work of the philosophers of science Thomas Kuhn and W. V. O. Quine (Murphy and Medin 1985: 308). Gopnik and Meltzhoff, for instance, cite Quine's "Two Dogmas of Empiricism," then point to the role of theory in his views about meaning (emphasis added): "Quine holds that meanings are deeply independent and deeply flexible. Meanings depend on larger theories of the world. Understanding the meaning of a term requires understanding the theory in which the term is embedded" (Gopnik and Meltzhoff 1997: 211). Advocates of the *theory theory* typically see this as a semantic holism that implies the appropriate unit of conceptual change is the *theory* that determines the meaning of concepts, not the concepts themselves.

Many discussions of Quine and the *theory theory* begin with his classic 1951 paper "Two Dogmas of Empiricism," where he challenged views he claimed were mere "dogmas." The first dogma is the traditional analytic–synthetic distinction that treats analytic claims exclusively as matters of meaning and definition, and synthetic claims as empirical and depending on facts about experience. Analytic truths, according to this "dogma," are true independent of all experience. Quine rejected this distinction on the grounds that analytic claims – including those of logic and math – are revisable in light of experience. Therefore there can be no determinate, unchanging definitions to give determinate and unchanging meanings to terms. The truth of analytic statements, like synthetic statements, is then dependent on both language and "extralinguistic fact" (Quine 1990: 34).

The problem with this analytic–synthetic distinction, according to Quine, is that it assumes that the meanings of terms are independent of each other. If that were true, the change in the meaning of one term would not affect the meanings of other terms. But, Quine argued, meanings are determined *holistically*, relative to a web of belief. They are products of complex relationships within a conceptual framework that is empirical at its periphery, but revisable throughout.

> The totality of our so-called knowledge or beliefs, from the most casual matters of geography and history to the profoundest laws of atomic physics or even pure mathematics and logic, is a man-made fabric which impinges on experience only along the edges ... A conflict with experience at the periphery occasions readjustments in the interior of the field. (Quine 1990: 37)

New observations or changes in a term's meaning result in changes throughout the framework, potentially even in the assumed truths of logic and math. Quine took this to imply that statements about the external world are not tested individually, but all together as a coherent whole.

In his 1962 *Structure of Scientific Revolutions*, Thomas Kuhn advocated a similar semantic holism. His inspiration originally came from Gestalt psychologists and the empirical work of the psychologist Jean Piaget, but he also credited the influence of Quine (Kuhn 1996: viii). Kuhn's holism is found most strikingly in the view that "paradigms" determine the meaning of scientific terms, and that scientific change occurs by a process analogous to political revolution, in the replacement of one paradigm by another. While he used the term *paradigm* in multiple ways, one prominent way was as a "disciplinary matrix" that contains a conceptual framework (Kuhn 1996: 200). A paradigm in this sense is the "entire global set of commitments shared by the members of a particular scientific community" (Kuhn 1977: xix).

What is important for purposes here is that the meanings of the terms that appear in the paradigm are determined at least in part by the implicit conceptual framework, and the other terms that function in the framework. Meaning is therefore holistic in the sense that concepts cannot be defined independently of other concepts in the framework. Kuhn thought this implied that those who work in different paradigms can be regarded as members of different linguistic communities. They use terms that are superficially the same, but have different meanings and refer to different things. The term *planet*, for instance, meant one thing in the Ptolemaic paradigm with its earth-centered model of the heavens, and another in the Copernican paradigm with its sun-centered model. In the former, the term *planet* referred to the sun. In the latter it did not. And consequently in the former paradigm, the term *sun* thereby had a different meaning than in the latter, since it was describable as a planet in the former but not the latter (Kuhn 1996: 128). Similarly, the term *water* meant one thing in the theory of Aristotle that contrasted it with the other material substances – earth, air and fire. But it has another meaning in modern chemistry, where it is composed of the elements oxygen and hydrogen, and is understood within the modern theory of the elements and chemical bonds.

According to Kuhn then, conceptual change occurs through a change in the paradigm – the overarching conceptual framework. Because all the meanings and references of terms are dependent on the paradigm, neither meaning nor reference can change atomistically – independently of the meanings of other terms and ways of referring. Kuhn took this to imply that conceptual change is discontinuous. The meanings of conceptual terms change when one paradigm is replaced by another in a scientific revolution. And they all change together. This, Kuhn argued, meant that theories in different paradigms are incommensurable. They are epistemically incommensurable in that, when paradigms change, the evaluative standards implicit in the paradigm change. So there are no unchanging paradigm-independent standards by which to evaluate the elements of different paradigms. But there is also a semantic or, more generally, a *linguistic incommensurability* in that the terms that appear in different paradigms do not have the same meaning in all the paradigms in which they appear, nor do they refer to the same things. The terms *planet* and *water* have different meanings and refer to different things in different paradigms.

Since the terms of scientific theories are often explicitly defined relative to other terms, this holistic approach is at some level uncontroversial. If the term *water*, for instance, is *defined* by reference to the terms *oxygen* and *hydrogen*, then the meaning of *water* is clearly tied to the meanings of these other terms. And their meanings are tied more generally into the overarching theory of the elements. So of course the meaning of concept terms within a theory is dependent on the meanings of other concept terms. Nonetheless, this holism has not gone unchallenged, most notably by an approach that gives priority to reference.

THE CAUSAL THEORY

The holism we see in Quine and Kuhn is a natural consequence of the Fregean descriptivist assumption that meaning determines reference. A concept term picks out something on the basis of a description that serves as the basis for its meaning. The description of water in terms of two hydrogen atoms and one oxygen atom, for instance, refers to whatever fits that description. So how a term refers and what it denotes are dependent on meaning, and changes in meaning. But there are good reasons to doubt that this is always the way meaning and reference work. First, we can often successfully refer to something without having anything like a good definition, or even a reasonable description. We can refer to something by ostension – by pointing at it. Or we can point at it indirectly with a clearly non-definitional description such as "the stuff in the glass." Second, at least sometimes the relation between reference and meaning may go the other way. Instead of meaning and description determining reference, reference might sometimes determine meaning. This is the idea behind a prominent *causal* approach to reference, developed by Saul Kripke and Hilary Putnam.

Kripke and Putnam agree with the Fregean approach in its commitment to the view that meanings are "not in the head." The meaning of a term is not to be associated with a mental entity of any sort, but is abstract and public, in the use of language (Kripke 1972; Putnam 1990: 308). But Kripke and Putnam disagree with the Fregean approach in its commitment to the view that description determines reference.

185

Rather, reference comes first, and then description and meaning. Or, in the terms preferred by Kripke and Putnam: "extension," the set of things denoted by a term, comes first, and the "intension" or meaning of the term is determined by the investigation of its extension.

The basic idea here is that there is some public "baptism" of a name or term, which gets used to refer to something by virtue of ostension – literal pointing at something, or by some descriptive phrase that serves to "point." The description "the stuff in that glass," for instance, points to a particular substance, whatever it is, and however it may *come to be described and defined*. Once reference has been fixed, the name or term and its reference gets passed on from speaker to speaker, who learn how to apply the term. This naming, however, is a social act and must be carried out by the appropriate sort of person in the appropriate context. There is a division of linguistic labor that reflects the more general division of labor, and that gives some speakers the authority to set reference in particular contexts, and determine the application of the term in future contexts. In scientific contexts, it is the scientist with the appropriate background for identifying the reference (or extension) of a term. This is true, even though others without the expertise may use the term. Putnam explains relative to the term *water*.

> [W]ith the increase of division of labor in the society and the rise of science, more and more words begin to exhibit this kind of division of labor. 'Water', for example, did not exhibit it at all prior to the rise of chemistry. Today it is obviously necessary for every speaker to be able to recognize water (reliably under normal conditions), and probably every speaker even knows the necessary and sufficient condition 'water is H_2O'; but only a few adult speakers could distinguish water from liquids which superficially resembled water. In case of doubt, other speakers would rely in the judgment of these 'expert' speakers. Thus the way of recognizing possessed by these 'expert' speakers is also, through them, possessed by the collective linguistic body, even though it is not possessed by each individual member of the body, and in this way the most recherché fact about water may become part of the social meaning of the word while being unknown to almost all speakers who acquire the word. (Putnam 1990: 311)

Reference, then, is historical in that it is tied to a particular baptism, and it is social in that the initial baptism is accomplished, and later usage is determined by the subclass of speakers who have the appropriate sort of expertise.

After the baptism the relevant linguistic community then finds out what the term means, by finding out about the nature of whatever the term refers to. We discover the meaning of the term *water* for instance, by investigating the nature of the stuff *water*, refers to. When we discover that water has a particular microstructure – H_2O – we have discovered something about the meaning of that term. We have discovered what water is and what a definition would include. On the other hand, there may be many properties that might help us operationally in the identification of water in the world. We might note that water is transparent and has a particular taste (or doesn't). These properties, however, don't tell us what water is in the same way its microstructure does.

Because reference is set by the initial baptism, a term is, according to Kripke, a "rigid designator." This means that it applies to the same things or same stuff, in "all possible worlds" (Kripke 1990: 290). Putnam agrees, arguing that, because these terms are "indexical" in the baptism, they have their reference set to whatever it was that was identified at the initial baptism. The "this stuff," pointed to in a particular context, sets the reference (Putnam 1990: 313). So the term or name applies whenever this stuff or this thing is encountered. This claim, however, demands an answer to the question: what does it mean to be the same thing or same stuff? The answer, according to Putnam, is that *sameness* is determined theoretically. Any stuff is the same stuff as what is referred to as *water* in the initial baptism, just in case it has the same microstructure – it is H_2O (Putnam 1990: 314). Meaning and reference, then, are determined by the social factors in the practice of science – the division of linguistic labor, and the contribution of the real world – what we find out about the reference, or rigid designation of the term.

REFERENTIAL VAGUENESS

The causal theory of reference has some advantages over the descriptivist tradition of Frege. It gives an account of how we can refer to things that we may know little about, or have false beliefs about, and therefore cannot give a correct description. But it has its problems as well. First, it is not clear that it can deal with the problem of co-referential terms that have different meaning (Soames 2005: 35). The terms *morning star* and *evening star* still seem to have different meanings, connoting different temporal location, even though they have a common reference. Second, its assumption of rigid designation does not seem to reflect actual practices. The history of science reveals that instead indeterminacy in reference is common, and that, as we find out more about what our terms refer to, we sometimes make adjustments in reference. And we make these adjustment on the basis of practical reason. In the actual practice of science, terms such as *water* and *species* do not seem to be the *rigid designators* assumed by Kripke and Putnam.

Putnam argued for rigid designation on the basis of what has since become a famous thought experiment.

> [W]e shall suppose that somewhere in the galaxy there is a planet we shall call Twin-Earth. Twin Earth is very much like Earth; in fact, people on Twin Earth even speak *English*. In fact, apart from the differences we shall specify in our science-fiction examples, the reader may suppose that Twin Earth is exactly like Earth. (Putnam 1990: 309)

The important difference between Twin Earth and Earth is the reference of the term *water*.

> One of the peculiarities of Twin Earth is that the liquid called 'water' is not H_2O but a different liquid whose formula is very long and complicated. I shall abbreviate this chemical formula simply as XYZ. I shall suppose that XYZ is indistinguishable from water at normal temperatures and pressures. In particular, it tastes like water and it quenches thirst like water. Also, I shall suppose that the oceans and lakes and seas of Twin Earth contain XYZ and not water, that it rains XYZ on Twin Earth and not water, etc. (Putnam 1990: 309)

Putnam claimed that, if travelers from Earth visited Twin Earth, those from Earth would think that *water* on Twin Earth meant the same thing it did on Earth. And if travelers from Twin Earth visited Earth, they would think that *water* means the same thing on Earth as it did on Twin Earth. These suppositions would be false, and would be corrected when the differences in structure were discovered. The English word *water* would have one meaning on Earth – *water*E, and another on Twin Earth – *water*TE. This is because there would be two different extensions determined by structure – the stuff that is H_2O versus the stuff that is XYZ (Putnam 1990: 309).

Like many thought experiments, Putnam's argument relies on intuitions about linguistic dispositions, what we *would* do if confronted by the assumed facts. Many philosophers endorse his intuition that the property of being H_2O trumps all other properties in determining sameness – appearance, taste and so on. The term *water* can therefore *only* refer to substances that have this particular microstructure. But this intuition is not unequivocally supported by actual cases with similar circumstances, as Joseph LaPorte has shown. The mineral jade, or "nephrite," has enduring significance in Chinese culture. It has been used to make scepters, coins, and jewelry, and has been highly valued in all forms. Because of this value, there have been numerous "jade imposters," substances that superficially look like jade but are not of the same chemical formula – $Ca_2(Mg,Fe)_5Si_8O_{22}(OH)_2$ – and do not work as well for the many traditional uses of jade (LaPorte 2004: 95). One particular imposter stone, first used near the end of the eighteenth century, has a complicated history. It is very similar in feel and hardness, but was discovered in 1863 to have a different chemical formula than the traditional jade – nephrite. This stone, known first as "kingfisher jade," has the formula $Na(Al(SiO_3)_2$ and came to be identified as "jadeite," distinguishing it from nephrite. Here seems to be a case similar to Putnam's XYZ of Twin Earth. If Putnam (and Kripke) were right, wouldn't we expect that, as soon as the structure – the chemical formula – of this false jade was discovered, it would cease to be regarded as jade?

> The question that faced the Chinese was whether this new jade, or kingfisher jade, was *"true* jade." Nephrite was unquestionably the "true jade" and had always been recognized as such. Jadeite was not so clear a case. The fact that "jade" ("yü") was used in compound expressions like "new jade" and "kingfisher jade" does not, of course, indicate that it was "false jade." This is especially clear because similar expressions refer to nephrite. If jadeite came to be called "new jade," nephrite came to be called "old jade." "Kingfisher jade" suggests a common color of jadeite, and similar expressions are given to nephrite of various colors: Different shades of white nephrite are called, for example "mutton fat jade" and "camphor jade." Instead, "kingfisher jade" is a term that had itself been applied to a green shade of nephrite from Turkestan before the term was applied to jadeite. (LaPorte 2004: 96)

What seems to have actually happened is that jadeite came to be commonly regarded as a true jade, even though its chemical composition is not the same as nephrite – the original "true jade" (LaPorte 2004: 97). This is in part due to its strong similarity to nephrite relative to practical use. Both nephrite and jadeite are very hard and carve in similar ways, and their appearance is similar (LaPorte 2004: 98–99). There was, it is true, an inclination at one time to restrict the term *jade* to nephrite, but today common usage "designates both nephrite and jadeite" (LaPorte 2004: 100). The initial baptism did *not* fix the reference of the term *jade* based on microstructure.

If this example seems problematic, there are other cases where the original baptism similarly does not seem to have fixed reference in the way supposed by Kripke and Putnam. The term *ruby*, for instance, has long been used for a red mineral of the formula Al_2O3. But it was later discovered that there are other stones of the same composition with various impurities that cause them to be different colors. On the Kripke-Putnam analysis we would expect that the term *ruby* would come to be used to apply to these stones as well. That did not happen. The term *ruby* continues to be reserved for the red *variety* of Al_2O3 (LaPorte 2004: 102).

Perhaps even more pertinent is the case of heavy water. In Putnam's Twin Earth thought experiment, when it was discovered that water on Twin Earth had a different microstructure, we were supposed to conclude that the term *water* meant one thing on Earth and another on Twin Earth, by virtue of the different microstructures of the Earth and Twin Earth substances. But it might have gone another way. We might have concluded that there are two varieties of *water*. Something like this seems to have happened in a real case from Earth. In 1931 it was discovered that there is a variety of hydrogen, a stable isotope occurring naturally, that has the standard nucleus with single proton, but also has a neutron. This variety, dubbed "deuterium," has the same atomic number as ordinary hydrogen, but with the additional neutron it has extra mass. When combined with oxygen in the familiar H_2O ratio, it forms "heavy water," or "deuterium oxide" (LaPorte 2004: 104–108).

What is important here is that water in general contains both varieties of hydrogen atoms, the more common lighter "protium," as well as the less common heavier deuterium. When it was discovered that water was H_2O, the reference or extension of the term *water* could have been fixed to denote only the light water. But since heavy water was also present (albeit in small quantities), it might have been fixed to denote only the heavy water. Or it might have been fixed to denote both, treating light water and heavy water as two varieties of the same substance. Or, as LaPorte argues, it might have just been indeterminate as to what stuff the term *water* referred to. "Given that both of these related, authentic, salient natural kinds were instantiated by the samples, no kind was seen to have an overpowering claim to have been the referent of *water*" (LaPorte 2004: 108).

THE RETURN OF THEORY

These apparent counter-examples raise an obvious question about the Kripke-Putnam claim that terms like *water* are rigid designators – they necessarily refer to the same thing or stuff they referred to in the original baptism. What does it mean to be "the *same* thing or stuff?" When we point to a particular sample of water, we don't by that ostensive act establish what it is about it that is relevant to the extension of the term – the important similarity or similarities it has with other things of the same kind. Putnam claims this is determined *theoretically* over time in an investigation: "the relation same$_L$ is a theoretical relation: whether something is or is not the same liquid as this may take an indeterminate amount of scientific investigation to determine" (Putnam 1990: 310). Further investigation may reveal facts that require a revision in the factors that determine sameness. When we thought that the important facts for the determination of the reference of *water* were its qualities, we were later proven wrong and facts about microstructure came instead to be relevant to sameness and reference. Further investigation may reveal yet another factor that is more important in determining reference – and use of that factor may well result in a change of the reference of *water*.

But there is more to this causal story than just the discovery of new facts. First and most obviously, reference might well change with theory change, and not merely through the discovery of new facts. When a theory changes, what counts as "the same" may change as well. There is therefore no guarantee that the set of things a term is assumed to refer to – the extension of the term – in the initial baptism, will not change when the relevant theory changes. This is most obvious relative to terms such as *planet* that meant one thing on the Ptolmaic model and another on the Copernican, and consequently referred to different things in each model. If theory determines sameness in reference, it is hard to see how description does and cannot play some role or other in determining reference.

Second, even in the absence of theory change, there may be multiple ways to think about sameness on the basis of that theory. Any given theory has many different ways to construe "the same" – perhaps on the basis of functioning, qualitative features, quantitative relations or causal factors. With the heavy water example, the relevant way to be "the same" might be understood in terms of molecular composition and atomic number: two hydrogen atoms (of any variety) and one oxygen (of any variety). In which case light water and heavy water would be the same in the relevant way, and all samples of both light water and heavy water would constitute the extension of the term *water*. But if "the same" were to be construed in terms of the atomic *structure* of the constituent atoms, then heavy water and light water would be different kinds of

191

things, and would not constitute the extension of the term *water*. Or we might decide on the basis of functioning in organic processes. Heavy water does not function in precisely the same way for organisms, so may not be "the same" in the relevant way. How do we decide which way to think about "the same" here?

What these examples suggest is that there may be an indeterminacy in both meaning and reference, where it is not clear precisely what we should include in the description associated with the meaning of a term or what that term refers to. At various points in time, there seemed to be no determinate facts about whether the term *jade* applied to nephrite, jadeite or both. This indeterminacy is not without its limits however. The term *jade* might potentially refer to one or both of these two similar stone, but not to minerals that are radically different in color, hardness, and microstructure. And the term *water* might potentially refer to *light water*, *heavy water* or both, but not to substances that differ greatly in qualities and microstructure. What these terms may potentially refer to depends on the context of the use, and the theoretical background. So even though there is some indeterminacy in the reference of these terms, that indeterminacy is constrained by what may be termed the *reference potential* of the term (Kitcher 1978; Burian 1985). The reference potential of a term is the range of reference (or extension) that is plausible, given a particular meaning of the term. Consequently, the more vagueness or ambiguity there is in the description of a term, the greater its indeterminacy and the broader the reference potential.

LaPorte claims that in cases such as *jade* and *water* the meaning of the term is just stipulated, so that it more clearly and unambiguously applies to the relevant things (LaPorte 2004: 118). This is part of scientific progress. Similarly, the term *species*, according to LaPorte, has been stipulated and clarified by evolutionary theory in just this way, and has therefore come to more successfully refer to the relevant things in nature (LaPorte 2004: 131). Whether LaPorte is right about scientific progress in general, and the species concept in particular, I shall set aside for later discussion. The important insight for present purposes is that the specific meaning and reference of a term is ultimately the product of a *decision* to use the term in a particular way. But to understand *why* a term gets used the way it does, and how its use changes, we need to say something about the reasons for these decisions about usage. In the next section we will return to the history of the species concept, as laid out in chapters 2 through 4, in order to understand the reasons behind particular ways of using the terms *eidos* and *species*.

EIDOS AND SPECIES

Implicit in the history of the species concepts laid out in earlier chapters is a set of assumptions about the nature and functioning of species concepts. Because we have been looking at the philosophical treatment of concepts in this chapter, we are now in a position to make many of these assumptions explicit. The first assumption is that species concepts are to be understood as Fregean. That means that there is both a "sense," or meaning, associated with the term *species* and a "nominatum," or reference. By assuming this, we need not assume that the other theories of concepts outlined at the beginning of this chapter are irrelevant. Species concepts may, or may not, have associated abilities and mental entities. The second assumption is that there is some definitional structure or other. The description constituting the meaning of a concept might have a classical structure – a set of necessary and sufficient definitional conditions, or a probabilistic cluster of conditions. In both cases, though, not all of the description is definitional. The third assumption comes from the *theory theory*, and is that, whatever definitional structure a concept has, the content and coherence of the associated description is due to some theory or other. The conditions associated with the description are related by some theoretical assumptions about how they are connected, and how and why they are important. The fourth assumption is that there is indeterminacy of both meaning and reference. The meaning and reference of a species concept may not be definitive, and they are always potentially revisable in light of its empirical application or of some theoretical consideration. The final assumption is that this indeterminacy is constrained. A concept term has a *reference potential* that indicates possible reference, but rules out other patterns of reference, on the basis of the corresponding vagueness in the meaning of the concept term.

In our return to the history of species concepts, we will also be sensitive to two other questions. First, what are the reasons that governed the decisions about meaning and reference? These reasons for decisions may be based on scientific values, practical goals and aims. or they may be grounded on broader cultural, non-scientific motivations. Second, what social factors are potentially relevant? This question is asking us to think about the training and methods of particular disciplines and approaches, as well as social structure – how science gets grouped into disciplines and sub-disciplines, and units of interaction. As we shall see, these factors are helpful in understanding the species problem. With these philosophical assumptions now explicit, we can more profitably return to the history of species concepts as laid out in chapters 2 through 5.

As we saw in chapter 2, Aristotle seemed to have used the term *eidos* in three distinct ways with three distinct meanings. The first meaning is relative to Aristotle's theory of the method of division, and can be understood as a logical universal. This is the meaning of the term as it was used in Aristotle's logical works, and reflects the role of an *eidos* within the method of division as a subdivision of a *genos*, based on the presence of differentia. On this sense of *eidos*, there is no distinctively biological significance, even though some of the terms such as *bird* and *man* are themselves applied to biological things. Rather the primary meaning and reference of the term *eidos* in this sense are linguistic, classificatory and hierarchical. Its reference is to linguistic entities – predicates and universals, not to biological entities – organisms or group of organisms.

The second use of the term is as enmattered form, and is derived from the older meaning of "appearance" or "what one sees." The meaning here is not based on the method of division, rather on a theory about perceptual experience and form. Nor is the reference of the term in this sense linguistic, but to something in the world – a shape or structure of a particular individual thing that can be seen. Once again, this is not a distinctively biological use, as its applicability extends beyond biology to other domains.

The third meaning of the term *eidos* in Aristotle's work is as "immanent principle of organization and vitality." This meaning of *eidos* is tied to the associated meaning of *genos* as the matter that contains the potential for different paths of development through a process of differentiation. The meaning and reference of the term here are neither linguistic nor perceptual. The *eidos* is that which explains organic activity – a certain developing pattern of life. This sense of *eidos* is the most distinctively biological.

If this account of Aristotle's use of the term *eidos* is right, then this term, as used by Aristotle, is systematically ambiguous. It has distinct meanings that are based on different theoretical frameworks, have different practical motivations, and that refer to different kinds of things. The sense or meaning of *eidos* in each case is at least partly determined by the theoretical context of its use. The logical universal sense is to be understood within Aristotle's theory of division. The enmattered form sense is to be understood at least partly within a theory of the forms, and the principle of organization sense is to be understood within a

194

functional-developmental framework. In each case, meaning is dependent in some way or other on the theoretical commitments that are presupposed in each context – about the functioning of language in the first case, about the nature of the forms in the second, and about functional-developmental processes in the third.

As we also saw in chapter 2, after Alexander of Aphrodisias, commentators focused on Aristotle's use of *eidos* as logical universal. This was due in part to the disappearance of Aristotle's biological works, but also to the growing preoccupation with his logical works, *The Categories* in particular, and its analysis of universals as general terms of predication. The other Aristotelian uses of *eidos* – as enmattered form and developmental principle – were simply not relevant to the discussions that followed, that were framed in terms of the Latin *species* and in the context of two new powerful influences, Neo-Platonism and Christian theology. Porphyry laid out the framework for thinking about species for next centuries in a series of questions.

> At present ... I shall refuse to say concerning genera and species whether they subsist or whether they are placed in the naked understandings alone or whether subsisting they are corporeal or incorporeal, and whether they are separated from sensibles or placed in sensibles and in accord with them. (Jones 1969: 186)

What is notable here is that Porphyry was arguing not for a determinate reference and meaning of the term *species*, but for a *reference potential* and a set of correlated meanings. *Species* can potentially refer to something mental, or to something intrinsic to bodily things, or to something separate. Boethius, like others of his time, accepted the implied assumption about reference potential, but responded to Porphyry based on his Neo-Platonism, arguing that species exist as part of a timeless reality. This view was compatible with the idea found in Augustine that species were ideas in the Divine understanding that guided the Creation. We can therefore understand nature by understanding particular universals – species and genera – and we can understand God by understanding the general nature of universals. As we also saw in chapter 2, the meanings of these terms then came to be of significance in theological debates about original sin and the nature of the Trinity.

The details of these theological debates are less important for understanding conceptual change than the following. First, even though the debates began with a single Aristotelian sense as logical universal, the term *species* nonetheless remained ambiguous. There were multiple meanings, and reference was not rigidly set to a particular kind of thing. Second, the reference potential was determined by several distinct theoretical factors – Aristotle's views about logic, Neo-Platonist metaphysics and Christian theology. Decisions about reference and meaning could potentially be based on any of these three factors, with their corresponding sets of values – logical, metaphysical and theological. One could serve the needs of logic, Neo-Platonic metaphysics or theological doctrine. Third, the definitional structure was largely determined by the Aristotelian logical foundation. Whatever else *species* were thought to be in this debate about universals, they were from the start linguistic entities – universal terms. This core commitment is surely due to the fact that the standard education began with Aristotle's *Categories* and thinking about species began within that linguistic and logical framework. Perhaps most important for the present discussion is that, from the disappearance of Aristotle's biological work to the Renaissance, there was virtually no role for biological theory. The species *man* and *animal* were in this linguistic starting point, equivalent to the species *justice* and *the good*. *Species* did not function within a theoretical biological framework or a fixed biological classification – because there was no established biological framework or classification. Finally, the practical goal of a theory of universals was to reconcile Plato and Aristotle within a theological framework. Any theory about species needed to preserve theological commitments. This would be particularly important because those who thought and wrote about the nature of species and genera during the period from Boethius through to the Renaissance typically had theological training. They were theologians and logicians, not biologists.

BIOLOGY AND THEOLOGY

In the third chapter, and with the medical herbalists of the early Renaissance, we saw the beginnings of a biological theoretical framework, a biological meaning to the term *species*, and a biological reference potential. The practical goal of the medical herbalists was the identification of plants to be used by the apothecaries in the formulation of medications. They started with the texts of the ancients, such as Theophrastus, Dioscorides and those newly rediscovered of Aristotle. But because the texts were incomplete, focused on the Mediterranean flora and were sometimes corrupted, the herbalists turned to the observation of nature for guidance. They actually looked at plants! The term *species* for them, then, had a medical sense, as a type of plant for use in medical treatment. This use was not biological in the modern sense however. There was no fixed hierarchy, and the terms *species* and *genus* were used at multiple levels. Nor was there any consensus about species and genus groupings. There was, therefore, an indeterminacy of reference, even though the reference potential was now more distinctively biological.

With the early naturalists from Cesalpino and John Ray to Linnaeus, there were two important theoretical developments. First was the development of a fixed, hierarchical biological classification. For Cesalpino and Ray, *species* and *genus* were the only taxonomic categories, although Ray began to distinguish subdivisions of the species category, and referred to some higher-level genera as "orders." Linnaeus developed and codified this classificatory framework whereby *species* came to occupy a single, basal taxonomic level. The term *species* then came to be understood at least partly through its place in the classificatory scheme – as the fundamental units of classification. And while this was primarily a biological classificatory system, it was not exclusively so. Ray recognized metal species and Linnaeus recognized mineral.

The second theoretical development was a synthesis of the Aristotelian assumption about the importance of functional traits, particularly nutritive and reproductive, with the theological assumption that species, whatever they are, were created by God, and thus represented ideas in the Divine understanding. This had two important consequences. First, there was a duty to understand nature so as to better understand God's plan. Second, it led to the first explicit theories of biological species. Most notable in the first theorizing about species is the emphasis on genealogy. Ray argued for a genealogical criterion: an organism was of a particular species if it was descended from organisms of that species. Linnaeus followed up on this idea and argued that species should be conceived in terms of a genealogical lineage stretching back to the original act of Divine creation, even though he identified that act later with the founding of a class or order. What is important in this development is that the theological framework was informing the meaning of the term *species* as it applied to the biological world. This in turn affected the definitional structure. The theological framework tells us that the place in a lineage is more important than mere similarity, because the lineage connects each organism back to the original acts of Divine creation. This turn to genealogy suggests that species change (for Linnaeus through hybridization) was possible.

Buffon, although he was not theologically motivated, accepted and developed this definitional structure that species were first and most fundamentally lineages. He argued that this fact further implied that species were *historical individuals*, and accepted the possibility of change in species lineages. He also focused on the tension between the abstract logical order in biological classification and the real order in nature. Notice this is opposed to the views of the medieval thinking that conceived the species problem *as a problem of universal terms and logic.* So even while there was referential indeterminacy here, it was not clear which lineages were species, the reference potential ruled out linguistic entities. Whatever else species were, they were lineages.

What is important in this period is the role of the theoretical frameworks, particularly the theological, in the meaning and definitional structure, as well as in the reference and reference potential of the term *species.* Species were understood as products of divine creation, and were constituted by the lineages that led back to the creation. They were *not* to be understood in terms of the language of predication, as universal terms. And while species were not exclusively biological, there was an increasing focus on the biological world. But nonetheless, and because of its theological significance, the idea that species were lineages came to be a core commitment in the understanding of the nature of species. It came to be part of the definitional core. Other factors, such as similarity, may have had descriptive significance, but were of less definitional significance than genealogy. Implicit in the definitional structure here also is the hierarchical pluralism that distinguishes theoretical and operational concepts. Theology determined that being a lineage was the most important factor in the definitional structure, and other factors such as similarity were of secondary and merely descriptive importance. This is the situation in which Darwin found himself. Naturalists used many criteria for identifying species, but all agreed that genealogy was most important. As Darwin concluded, after considering the many different ways that naturalists had characterized species: "With species in a state of nature, every naturalist has in fact brought descent into his classification" (Darwin 1964: 424).

If the meaning and definitional structure of scientific concepts are theoretically determined, then we should expect the acceptance of Darwin's theory of evolution to have revolutionary implications for both the meaning and reference of *species.* So it may be surprising that, as we saw in chapter 4, he seems to have adopted the core definitional meaning of species as lineages from his theologically inclined predecessors.

There are obvious changes nonetheless. First, these species lineages, for Darwin, were to be understood as the branches of the evolutionary tree, as represented in his tree diagram from the *Origin*. Based on the fact that this was a process of branching and divergence, species lineages came to be understood as things that originate from mere varieties in the branching and divergence of evolution. This is clear in a passage quoted in chapter 4:

> I look at varieties which are in any degree more distinct and permanent, as steps leading to more strongly marked and permanent varieties; and at these latter, as leading to sub-species, and to species ... Hence I believe a well-marked variety may justly be called an incipient species. (Darwin 1964: 51–2)

One important difference between Darwin's conception of species lineages and those of Linnaeus and Ray is that earlier views *allowed* for change, but change was not part of the definitional core. For Darwin, it was a core, definitional condition of species lineages that they change.

With Darwin we also see the beginnings of an emphasis on the population aspect of lineages. This is most obvious in his "horizontal" distinction between varieties and species. Varieties were groups of organisms that were largely similar, but not distinct from each other. Species were groups of organisms that were distinct from other groups. At the same time, there were other relevant conditions to being a species, but that were more peripheral. Reproduction and sterility, ecological functioning, geographic distribution and similarity all were relevant, but not necessarily part of the definitional core. These factors may be best conceived in terms of the cluster models of definitional structure, and constrained to the descriptive periphery. If so, then in Darwin's conception of species, there was a definitional core that defines species as lineages, and a cluster of peripheral descriptive conditions based on morphology, reproduction, etc.

This sort of a definitional structure could give the appearance of systematic ambiguity in the term *species*. Species could be described as reproductively isolated populations, geographic populations or groups of similar organisms. If the descriptive periphery consists of disjunctive clusters of conditions, then there are multiple descriptions that we could potentially associate with the term *species* that have different reference potentials. What the term *species* refers to will depend on which conditions are included in the particular cluster of descriptive conditions not part of the definitional core. But *species* is also referentially indeterminate in part because populations and lineages have vague boundaries. As Darwin worried, there are multiple ways to divide lineages and populations, with no clear boundary demarking one segment of a lineage from another. And there are multiple ways to draw boundaries between varieties and species at a particular time. But this indeterminism is also constrained. Darwinian theory tells us what kinds of things *species* might potentially refer to – population lineages that diverge to some degree – even if it cannot provide unambiguous guidance in all instances.

In the period since Darwin, evolutionary theory has developed and changed in many ways. Most relevant for purposes here is the development of models of genealogical lineages and how they might be structured (see de Queiroz 1999). This has obvious relevance for any use of *species* that has as a core condition the requirement that they be lineages. There has also been much theorizing about the processes that generate and eliminate species, from the sympatric and allopatric models of speciation, the development of reproductive barriers, to the models of macroevolution that have introduced new theories of mass extinction. This theorizing has potential relevance in that speciation and extinction events are the endpoints of lineages, and therefore have significance in terms of the meaning and application of *species*. The development of new classificatory systems, cladistics in particular, also has significance insofar as it gives theoretical guidance about the categories and units of classification, starting at the base with species and extending upward. If there is a definitional core that is based on the condition that species are genealogical lineages, all of this theoretical development will have relevance to the meaning, and perhaps to the reference of *species*.

There has also been development of new technologies. Because it is now possible to analyze the genetic makeup of organisms, it is relatively easy to compare genotypes and establish gene frequencies. We therefore know much more about the molecular variability within and among populations, as well as over time. We have new technologies that allow us to identify and study signaling systems associated with mate recognition. We know much more about development and behavior, and how that might affect a variety of processes relevant to population lineages. In short we have at hand all the theoretical and technological resources that have seemingly resulted in the proliferation of species concepts we outlined in chapter 4, based on morphology, genetic similarity, reproduction and mate recognition systems, geographic isolation and more. Some of this technological development may be relevant to the definitional core of *species*, but much seems to be most relevant to the definitional periphery – the merely descriptive conditions that cluster around the core.

If the concept and term *species* has a definitional core – a set of conditions most important definitionally – and a set of disjunctive and merely descriptive peripheral conditions, the focus on different sets of peripheral conditions – interbreeding, morphology and so on – does not necessarily imply that different species concepts are in use. One way to think about this is in terms of vagueness in the definitional structure. There may be multiple, different subsets of conditions that can be satisfied to count for something to count as a species. Each subset would have the requisite definitional core – that species are segments of population lineages – and some different subset of the merely descriptive periphery – perhaps related to interbreeding, gene pools and morphological similarity. The particular use of the term *species* may or may not include particular descriptive conditions such as interbreeding. For sexually reproducing species this descriptive condition is relevant in a way it is not with asexually reproducing species. Since it is not clear a priori precisely what subsets of conditions are sufficient for a group of genealogically connected organism to be a species lineage, there is a vagueness in the meaning of the term *species*.

One virtue of this vagueness is that it allows for the indeterminacy of reference we sometimes see in science, as illustrated by the examples of jade, ruby and water. Another virtue is that it shows how theoretical and technological changes can sometimes be accommodated by changes in the descriptive periphery, without requiring changes in the definitional core. Conversely, it shows how, on the other hand, there can also be a change in the definitional core. This is a decision that will potentially be made on a variety of factors. In the cases of jade and ruby the factors were at least partly practical, based on the use of these minerals in everyday life. In the case of heavy water, it may also be practical, given that there are differences in the biological use of light and heavy water. Some organisms that require and use light water cannot use heavy water in the same way. This vagueness is particularly useful in the application of the species concept. As we discover new facts about biodiversity, we must often accommodate them within the theoretical framework in use. Some of these new facts will challenge the definitional core of *species*. But some will not, affecting only the descriptive periphery. We can now distinguish these two cases – one requires genuine theoretical change, the other does not.

THE DIVISION OF LINGUISTIC LABOR AND THE DEMIC
STRUCTURE OF SCIENCE

Hand in hand with the technological and theoretical developments of the past century are the increasing specialization and diversification of science. In Darwin's time, it was common for those we would consider "scientists" to work across broad areas of science. Darwin's own work covered a wide range from evolutionary theory to taxonomy, development, ethology, heredity and even geology. But after Darwin, there was increasing specialization into fields such as embryology, cytology, genetics, population genetics, behavioral biology, ecology and systematics. There has also been specialization within these new fields. In systematics, for instance, there is specialization according to both the organisms studied and the type of data used – morphological, molecular and behavioral. This specialization and diversification of science has had an important influence on the use of the term *species*. We can understand this effect by the consideration of two ideas, that science has a "demic" structure, and that there is a division of linguistic labor.

David Hull has argued that we should think about the social structure of science at least partly in terms of "demes" (Hull 2001: 178–179). Demes are subpopulations of organisms that directly interact with each other. In a given population, organisms will typically interact with some members but not others. Most obviously they might mate with nearby organisms, but not those further away. Analogously, scientists don't typically interact with everyone in the same discipline. Biologists don't typically interact with all biologists. Rather scientists interact only with those in a particular specialty or subspecialty. Molecular systematists who study fish tend to interact with other molecular systematists who study fish.

In general, authority is parceled out on the basis of this social structure. Those who work within particular disciplines, sub-disciplines and demes are usually given authority over the domain of that discipline, sub-discipline or deme. Among other things, they have linguistic authority over how terms are used, or should be used. Geneticists have linguistic authority over the use of the term *gene* for instance. This is the idea, advanced by Putnam and noted earlier in this chapter, of the division of linguistic labor. While Putnam focused on how reference was set – some speakers have authority to set reference – there is a division of linguistic labor about meaning as well. Some speakers have authority over the meaning and definitional structure of terms. Just as geneticists have authority over how the term gene is applied, they have authority over its meaning and definitional structure. So while the term *gene* has wide usage in various scientific disciplines and sub-disciplines, as well as the general public discourse, that usage is always subject to correction by those with linguistic authority.

This social structure and parceling of linguistic authority has played a role in the persistence of the species problem. The term *species* has, at various times, been within the proper linguistic domain of different disciplines. In the period after Aristotle and before the Renaissance, the term was used to refer to linguistic entities – universal or general terms, and those with linguistic authority were typically logicians and theologians. Beginning with Cesalpino and through to Darwin, the linguistic authority resided with naturalists in general, who typically worked within a theological framework. But after Darwin and with the increasing specialization of science, the linguistic authority became fragmented. Each discipline, sub-discipline and deme developed its own unique technical vocabulary over which it had linguistic authority. But some terms, such as *species*, continued to be used across many disciplines, sub-disciplines and demes, whether or not they have primary linguistic authority.

The details of usage and linguistic authority are beyond the main concern here, but there are some broad generalizations that may be helpful in understanding the species problem. The various disciplines that use the term *species* may each have *some* authority relative to the reference and meaning of this term, but *primary* linguistic authority seems to reside with systematists and evolutionary theorists. This is because species are the fundamental units of classification and evolution. Nonetheless, other sub-disciplines and demes have some authority over the usage of the term within their respective domains. Geneticists have authority over the application of the term *species* with regard to genetic similarities and differences. Ecologists have authority over the term relative to ecosystems and niches. But in such disciplines and sub-disciplines, the focus is often on the descriptive periphery that is relevant to that specialty – not the core definitional conditions. As each specialty focuses on that part of the descriptive periphery that is most relevant to its own subject, it may seem that there is a systematic ambiguity in the term *species*. It may seem to have different meanings and refer to different things. This is the heart of the species problem. The use of different species concepts that divide biodiversity in different and incompatible ways, by different disciplines and sub-disciplines, is perhaps at bottom better understood as a focus on different elements of the descriptive periphery that, *taken by themselves and without regard to the definitional core*, divide up biodiversity in different ways.

One theme of this book has been that the species problem is partly due to a misunderstanding of how species concepts function in the broader conceptual framework, and by neglect of the division of conceptual labor. The theme of this chapter is that the species problem is also due to a misunderstanding of how the term *species* gets used – how it comes to have meaning, the definitional structure of that meaning, and how it refers to things in the world. If we want to understand the species problem, we need to understand the feedback between meaning and reference, the definitional structure, the role of vagueness and referential indeterminacy, as well as the functioning of the division of linguistic labor within the social structure of science. And if we want to understand how species concepts change, we must also look at these same factors. What may be most surprising here is the value of vagueness and referential indeterminacy. The term *species* is vague in that it is not clear a priori what conditions are in the descriptive periphery, and which of those are relevant in any particular case. This allows for the revision of the meaning of the term in light of new empirical and technological developments that in turn allows us to continue to use the term *species* when our understanding of biodiversity and evolutionary processes change. It also allows use of this term by those with different theoretical and disciplinary interests. Nonetheless, these different interests do not force us into a linguistic incommensurability. When geneticists and systematists use the term *species*, they may mean something different by virtue of theoretical interests, emphasis and training, but they can still be speaking of the same thing in its definitional core – the species that also interest ecologists and paleontologists. To understand the species problem we need to understand this fact.

8

Conclusion

The main goal of this book is a comprehensive philosophical understanding of the species problem: the use of multiple and inconsistent species concepts that group and divide biodiversity in conflicting ways. In service of this goal, I have engaged in two projects. The first project, carried out in the first four chapters, is descriptive, and aimed at understanding the many different ways of thinking about species from Aristotle to modern biological systematics. What these chapters reveal is that, contrary to the Essentialism Story, there were many species concepts in use from Aristotle on, based on logic, theology, morphology, biological process, genealogy and more, that persisted through the Darwinian revolution to the thinking of contemporary biologists and systematists. There was no single predominant essentialist concept before Darwin, and no single predominant evolutionary concept in use after Darwin. Nonetheless, the influence of evolutionary thinking on species concepts is undeniable. Whatever else species are, evolutionary theory tells us they have beginnings, endings, cohesion, and they change over time.

The second project, carried out in chapters 5 through 7, is primarily prescriptive, aimed at understanding how we might better think about species. The main goal here is to see how our ways of thinking about species concepts can be clarified and corrected by an understanding first, of the conceptual framework in which species concepts function; second, the basic metaphysical status of species taxa; and third, the functioning of the species concepts themselves. The analysis in this prescriptive project is based on, first, the *division of conceptual labor* that distinguishes theoretical and operational species concepts; second, the *species-as-individuals* thesis, as the metaphysical stance that best coheres with the

205

evolutionary assumption that species are segments of population lineages; and third, an analysis of meaning and reference of the term *species* in terms of definitional structure and reference potential. To better understand the arguments presented here, a brief recap of the previous chapters is in order.

An important part of understanding the species problem is through its practical implications. This is the topic of the first chapter. There we saw how the adoption of different species concepts affects the way organisms are grouped into species taxa. One researcher may use a concept based on potential or actual reproduction, which groups organisms into species taxa in one way and results in one species count. Another might use a species concept based on morphological or genetic similarity, which groups organisms another way and with a different species count. Yet a third person may use any one of the other twenty plus concepts and arrive at an entirely different grouping and species count. This use of different species concepts that results in different groupings and species counts is common, and in some cases striking. As we saw in a survey of the research cited in chapter 1, the use of the phylogenetic species concept increased species counts by an average of 48.7percent over counts based on other concepts. This has real world significance. Most obviously, it affects how endangered species legislation gets applied. The concepts that result in greater species counts tend to make more organisms subject to endangered species laws – requiring sometimes costly preservation efforts. But independent of legislation, efforts to preserve biodiversity require that we get our understanding of biodiversity right. That requires we get species groupings and counts right. The problems outlined in this chapter suggest that this has not yet happened.

We can also understand the species problem through its history. This is the topic of chapters 2 through 4. In chapter 2, we began with the "Essentialism Story," the standard philosophical history of species concepts that assumes pre-Darwinian species concepts were predominantly essentialist, and that species have essences that make them timeless, unchanging and discrete. According to this story, the origins of essentialism are in Aristotle's method of logical division, a method that was adopted by many if not most of those who followed, including and especially Linnaeus, and up until Darwin. Darwin, according to this story, vanquished essentialist conceptions of species with his population thinking which emphasizes the variability within a population, and gradualism, which emphasizes the gradual change in populations over time. Population thinking and gradualism imply that there can be no set of unchanging essential, or necessary and sufficient, properties associated with a species taxon. The virtues of the Essentialism Story are its simplicity, dramatic power and rhetorical value. The problem with this story is that it is largely false. A closer look at the history of species concepts reveals the full complexity of the species problem, a complexity ignored by the Essentialism Story.

The flaws of this story are apparent when we look at a starring character in the Essentialism Story, Aristotle, and his use of the term *eidos* (later translated into the Latin *species*). He used *eidos* in three different ways – as *logical universal, enmattered form* and *developmental principle*. The first use, as logical universal, fits into his method of division where the *eidos* was a subdivision of the *genos*, at any level and with respect to any sort of division, biological and otherwise. But we also saw that Aristotle did not think it was possible to use this method of division to arrive at animal kinds. This conflicts with the Essentialism Story, which placed his method of division at the center of species groupings. Aristotle's use of division in the biological realm was primarily to establish the explanatory relations among the *parts* of animals, not the groupings themselves, which he largely just assumed.

Those who followed Aristotle adopted his use of *eidos* as logical universal, and largely ignored, or were unaware of, his other uses as enmattered form and developmental principle. They read only his logical works, primarily *The Categories*, and treated *eidos*, and its Latin translation *species*, as linguistic entities – as universal terms. From Porphyry and Boethius through to the early Renaissance, the emphasis was on the logic and reference of universal terms, whether they referred to something that subsists in things or were in the understanding alone, whether they were corporeal or incorporeal, and whether or not they were separate from sensibles. From Porphyry on, it was largely assumed that Aristotle's account was consistent with Plato's views. And after Boethius, one compelling demand was to render Aristotle consistent with Christian theology. In the emphasis on the logical universal sense of *species* and the efforts to reconcile Aristotle with Plato and Christian theology, there was little interest in the biological realm. The term *species* neither had a distinctly biological sense nor was it biologically informed. And as with Aristotle, that term had no fixed place in a classificatory hierarchy.

As we saw in chapter 3, and with the medical herbalists, there was a biological turn in the thinking about species. But the motivation was primarily pragmatic – the identification and classification of plants for use in formulating medications. The emphasis on observation and biological features, however, led to the first modern naturalists, Cesalpino, Ray and Linnaeus. Here we see the development of a fixed hierarchy, with species occupying the lowest level of the hierarchy. We also see the first explicitly biological theories of the species category. But what is striking, and contrary to the Essentialism Story, these thinkers, motivated by theological concerns, emphasized the genealogical aspect of species, rather than the morphological. An individual organism was a member of a species if it was descended from a member of that species, not on the basis of its possession of a set of essential, or necessary and sufficient properties. Morphological similarity was relevant, as offspring tended to resemble parents, but it was not determinative. The reasoning here was at least partly theological. Species were conceived in terms of a lineage descended from the original act of Divine creation. Notably, the alleged arch essentialist, Linnaeus, came to believe in the creation of new species by hybridization, from the originally created order or class. Buffon as well endorsed the idea of species as lineages, although he disagreed with Linnaeus about the possibility and value of a formal system of classification.

DARWINIAN EVOLUTION

This is the situation that confronted Darwin in his thinking about species. Genealogy and lineage were largely accepted as primary species criteria, but morphological similarity and distinctnesss, interbreeding and sterility, as well as geographic location were all taken to be relevant to species groupings and counts. By treating these other criteria as constitutive of species, however, it *seemed* that there were multiple species concepts that divided biodiversity in inconsistent ways. Because of this use of different species concepts, and because of the vagueness in "horizontal" species boundaries, Darwin expressed some skepticism about the reality of species. But as we saw in chapter 4, based on his principle of divergence that conceived species "vertically" in terms of lineages diverging and branching from the stem species of an evolutionary tree, he seemed to recognize the reality of species taxa historically. Species are those lineages that have passed through sufficient divergent change to become distinct and permanent. Species were therefore to be distinguished from varieties, which were less distinct and permanent.

Also in chapter 4, we saw how the development of Darwin's evolutionary theory affected species thinking: first in the discovery of the gene, and mutationist theories of species, and then in the development of the statistical techniques of the biometricians and the population genetics of the Modern Synthesis. Here the ideas about species that circulated in Darwin's time (and before) were developed, quantified and extended. Dobzhansky, for instance, focused on the idea of species as historical lineages and the processes that produce divergence: "Species is a stage in a process, not a static unit" (Dobzhansky 1937: 312). One important process recognized by Darwin, and emphasized by Dobzhansky, is the development of reproductive isolation. Julian Huxley and Ernst Mayr followed Dobzhansky in emphasizing the historical aspect of species, and the processes that function in speciation, reproductive isolation in particular. Mayr's advocacy of the *biological species concept*, in its multiple versions, is perhaps the most enduring legacy of this period – especially for all those who learn in their introductory biology courses that species are interbreeding or potentially interbreeding groups of organisms. G. G. Simpson followed Dobzhansky, Huxley and Mayr in conceiving species historically, or diachronically, and recognizing the processes involved in speciation, but he more explicitly focused on the priority of the historical aspect in his *evolutionary species concept*, that conceives species as a lineage "evolving separately from others and with its own unitary evolutionary role and tendencies" (Simpson 1961: 153).

After the Synthesis, there was an explosion of empirical information about biodiversity and the processes that produced it. There was also seemingly an explosion of species concepts, resulting in the twenty-plus species concepts identified and analyzed by Mayden. This led to the species problem and its disparate species counts, in its modern form and as described in chapter 1. This proliferation was in part due to an increasing dissatisfaction with the biological species concepts of Mayr. After all, an interbreeding criterion can only apply to organisms that interbreed sexually. Much of biodiversity does not do so. But the proliferation of species concepts was also due to the technological advances that helped in determining genotypes, understanding the chemical signaling in reproductive behavior, and more. Finally, the development of new systems of classification, cladistics in particular, led to new ways of thinking about the basal level of classification – species. Here we see the introduction of cladistic species concepts, such as the various versions of the *phylogenetic species concept*.

PLURALISMS

Part of understanding the species problem consists in understanding the philosophical responses to this problem. A simple monism, based on the acceptance of one species concept and the rejection of all others, seems unpromising. There is little chance that one species concept of the many in use will *replace* the others. A second, skeptical response is the conclusion that there are no *real* species things. This response, however, is biologically implausible. Members of species in general seem to have something in common that goes beyond incidental similarity or pragmatic grouping. That Socrates and Plato are both humans is a highly significant fact, and not a mere accident of linguistic practice. A third response, pluralism, is more plausible. With all of these competing species concepts, and no obvious reason to prefer one concept over all others, then perhaps there are species, but they are not all the same kinds of things. In chapter 5, we began with a look at various versions of pluralisms.

The first version of pluralism we looked at was the pragmatic pluralism of Kitcher, dubbed by him "pluralistic realism." According to this version of pluralism, there are multiple species concepts because researchers have multiple interests and there are multiple kinds of investigations. While this approach has the advantage of reflecting the many ways biologists have thought about species in actual research, it has the disadvantage that these inconsistent species groupings cannot all coexist within a single classification. Nor can there be a single way to count species for the application of endangered species law. If we *must* decide on a single species grouping, it looks to be possible only on political grounds – we simply favor the species taxa of one researcher over the others.

A second version of pluralism is ontological, and asserts that there are different kinds of species things, because there are different kinds of groupings in nature. Ereshefsky, for instance, argues that species are all *lineages*, but different kinds of lineages – interbreeding, ecological and monophyletic. This idea is perhaps implicit in a third kind of pluralism, the hierarchical pluralism of Mayden and de Queiroz that begins with the idea that there are two kinds of species concepts: theoretical concepts, which tell us what kinds of things species are, and operational concepts, which tell us how to identify species taxa. Because of this difference in functioning, theoretical concepts are evaluated on different grounds than the operational. A satisfactory theoretical species concept must be universal in application, and theoretically significant – given that evolutionary theory tells us species are the fundamental units of evolution. Operational concepts, on the other hand, must be applicable and reflect the empirical differences among species. What is important is that we can seek a single unified theoretical concept, while also accepting the inevitability of multiple, inconsistent operational concepts. We can accept that species are all segments of population lineages, without thereby rejecting morphological similarity, reproductive isolation, ecological functioning, and so on, as relevant to species groupings.

We can see how this hierarchical pluralism can preserve the standard and widespread idea that science involves unification. We can do so through the application of Whewell's model of consilience, whereby theoretical concepts unify phenomena through two processes, the "colligation of facts" and the "explication of conceptions," which in successful cases leads to the "consilience of inductions." Operational concepts, on the other hand, function in the application of theoretical concepts to phenomena, the colligation of facts, and are instead governed by a principle of proliferation – the more ways to connect theoretical concepts to phenomena, the better. In chapter 5, I proposed that we should therefore think of operational concepts in Carnap's terms, as "correspondence rules." This may prevent the confusion of theoretical and operational concepts that lurks behind the species problem.

On the basis of all this, I offered the following *division of conceptual labor solution* to the species problem. There are different kinds of species concepts, theoretical and operational, that are to be evaluated on different grounds. The best theoretical concept is the most universal and theoretically significant – the concept that recognizes two aspects of species – the synchronic (at a particular time), and the diachronic (over time). The idea advocated by both Mayden and de Queiroz, that species are segments of population lineages, is a plausible candidate. The many operational concepts, or correspondence rules, are ways to identify and individuate segments of population lineages. And as we gain empirical knowledge about morphology, genetics, reproduction, ecology, biogeography, we can identify species taxa in more ways. There is therefore a *principle of proliferation* for operational concepts, or correspondence rules – the more the better. The advantages of this solution are, first, that it explains the many concepts in use; second, it preserves the significance of *species* to evolutionary theory and biological taxonomy; third, it provides an explanation of the species problem in the confusion about the conceptual framework in which species concepts function.

In chapter 6, we addressed an associated debate about the metaphysical status of species as *sets* or *individuals*. We began with a brief sketch of the different ways we might approach metaphysics, including a naturalistic approach that sees metaphysics as continuous with science, and is aimed at understanding the basic and most fundamental things assumed by science. If we adopt that stance then we can ask the question: given that species taxa are population lineages, what more basic and fundamental sorts of things are they? In the Essentialism Story, the answer to this question is that species are natural kinds with essences. This way of thinking about species treats them as *sets*, things that are grouped together on the basis of possession of a set of essential properties. On the classic conception of natural kinds, the set of essential properties is conjunctive in that a conjunction of properties is necessary and sufficient to be a thing. There is also, however, a disjunctive essentialism based on a clustering of properties. According to this cluster essentialism, some subset of properties or other is necessary and sufficient, but not the same subset in all instances. The main problem with both of these approaches is that they are trying to conceive historical, changing things – species lineages – within an ahistorical and unchanging framework. We can avoid this problem by thinking about species either as spatio-temporal *individuals,* or as *historical kinds*.

The *species-as-individuals* thesis treats species as spatio-temporally restricted, concrete individuals, analogous to individual organisms. Individual organisms are then *parts* of a species individual, rather than *members* of a set. Like organisms, species taxa have beginnings and endings, with change in between, and a sort of cohesion. According to Ghiselin and Hull, the most prominent advocates of this thesis, individuality is not exclusive to species taxa or individual organisms, but extends throughout nature and to social entities. And what is important in individuality is the historical connection and the process of cohesion. While it is roundly agreed that members – or parts – of a species are connected genealogically, it is often doubted that there is a comparable degree of cohesion between members of a species and parts of an individual organism. When the organs of an organism are separated, that organism dies, because it is so tightly integrated. But when the parts of a species are separated, there is typically no such catastrophic outcome. As we saw in chapter 6, Hull and Ghiselin responded by pointing out that the cohesion of individual organisms varies throughout nature, from the highly integrated vertebrates to the much less integrated social organisms. Similarly, there are often unnoticed factors of integration and cohesion among species.

Another approach that treats species historically is the view that they are *historical natural kinds*. On this approach, species have essences, but these essences are not the intrinsic properties employed by traditional or cluster essentialists. They are instead the historical relational properties that determine where organisms are within a genealogical nexus. An individual organism is a member of a species because of its genealogical relation with other things in that species. While this historical essentialism is superior to the traditional and cluster essentialism, in that it is historical and more consistent with evolution, it gives us little guidance in thinking about the historical nature of species, telling us *only* that there are these historical relations. But all organisms are part of the same genealogical nexus, and thus are connected historically, even though not all organisms are part of a particular species. The conception of species as historical kinds, therefore, gives us little guidance in working out the particular historical relations that are relevant. The real advantage of the *species-as-individuals* thesis is that it gives us guidance for thinking about the important factors that generate new species lineages, and that operate in the birth and death of species. It also gives us guidance in how to think about change over time, and the cohesion that makes a group of organisms a species. So while we can talk about species as either sets or individuals, there are great heuristic advantages in thinking about them as individuals.

In chapter 7, we returned to species concepts, but instead of looking at the conceptual framework in which they function, we looked at the functioning of individual concepts themselves. To do so, we began with some standard philosophical assumptions about the nature and functioning of concepts. One assumption is that concepts are "Fregean," and have a meaning ("sense"), and a reference ("nominatum"), and that the meaning determines the reference. Another assumption is that the meaning of a concept is to be associated with a description, and that this description has a definitional structure. There might be a classical structure, whereby there is a definitional core, consisting of a set of necessary and sufficient conditions, and a descriptive periphery that is, strictly speaking, not definitional. Or there might be a clustered definitional structure whereby the definitional core is some subset or other of a set of descriptive conditions. A third assumption, the *theory theory*, is that the definitional core is given content and coherence by some theory. This last assumption seems to imply some sort of holism – in that the meanings of concept terms depend on how the concept fits into an overarching theoretical framework.

We then took a look at a competing approach, the *causal theory*, which proposes meaning is determined by reference, which occurs through a baptism by ostension. The meaning of the term is then determined by an empirical investigation into the kind of thing referred to in the baptism. This is allegedly an atomistic account of meaning, and is supposed to avoid some of the worries raised about the holism of the descriptivist accounts. But, as we saw in this chapter, theory intrudes here as well, in the determination of the sameness of reference. We need to have a theory to tell us when we are pointing to the same kinds of things. Because of this, the reference of a term is often vague, and not rigidly designated. We saw this in the use of the terms, *jade*, *ruby* and *water*. In these cases, there is a referential indeterminacy, constrained by a reference potential, and with a corresponding vagueness of meaning. How indeterminacy and vagueness are resolved depends on decisions affected by theoretical and practical considerations.

In this chapter we then returned to the history of the species concepts to apply these insights about concepts: first, the role of theory in generating descriptive content and definitional structure, second, the referential indeterminacy and reference potential of the use; third, the social and practical factors that played a role in decisions about meaning and reference. What was also important was the "demic" social structure of science and the division of linguistic labor. Each sub-discipline of the biological sciences has its own theoretical interests and linguistic practices that reflect those interests. One implication is that researchers tend to focus on the part of the definitional structure that is most relevant – even if it is in the descriptive periphery. By doing so, researchers from different sub-disciplines can think about species, and understand the term *species* in different ways, by focusing on different aspects of the descriptive periphery. In this way, it is possible to accept the same definitional core of *species* while still focusing on different aspects of the descriptive periphery.

THE SOLUTION TO THE SPECIES PROBLEM

We are now in a position to pull together several threads in understanding the solution to the species problem proposed here. The first thread, from chapter 5, is the idea of the division of conceptual labor, based on the distinction between theoretical species concepts, that tell us what kinds of things species are, and operational concepts that tell us how to identify and individuate species taxa. Theoretical species concepts are grounded on evolutionary theory that treats species as the fundamental units of evolution. They are the things that originate in speciation, change over time, and go extinct, with a synchronic, population dimension, and a diachronic, lineage dimension. Operational concepts – or correspondence rules – are dependent on what we discover in empirical investigation into biodiversity. Because there is great variability among organisms, we must use different criteria for identifying and individuating the species of different organisms. And in general, those with different theoretical interests, who study different traits, processes and organisms, will focus on different operational criteria and different correspondence rules. By distinguishing these correspondence rules, which are sensitive to practical interests, from the theoretical concepts that are not, we can avoid the confusion that results from treating all species concepts as equivalent, competing ways for understanding the nature of species.

The second thread, from chapter 7, is based on the idea that species terms have a "sense" or meaning associated with a descriptive content, and a "nominatum" or reference. The descriptive content of the term *species* has a definitional core determined by evolutionary theory, and a non-definitional, descriptive periphery. In chapter 5, I endorsed the definitional core of *species* advocated by Mayden and de Queiroz, as segments of population lineages. Whereas the definitional core is primarily dependent on the theory in which the term *species* functions, the descriptive periphery is largely discovered by empirical investigation into the features associated with the reference potential – those things that might plausibly be regarded as *species*. Among those features we discover, are process features such as reproductive tendencies and ecological functioning, various types of similarities, historical relations and more. Particular features in the descriptive periphery will be relevant to some organisms but not others. Potential interbreeding, for instance, will be relevant to sexually reproducing organisms, but not to those that reproduce asexually. The descriptive periphery will therefore have a disjunctive structure whereby some subset of the descriptive content will apply to some organisms, while another subset may apply to other organisms.

Which subset of descriptive features is relevant in any particular case will depend on the division of linguistic labor and the *demes* of science. Science is structured so that scientists work together and interact in smaller groups. Each group will have particular research interests and a way of using terms such as *species*. Within each deme there will be

215

linguistic practices unique to that deme, practices that reflect the relevant theoretical interests, and are reflected in the particular descriptive content, the descriptive periphery in particular. So while it is true that those who study one group of organisms may use the term *species* differently than those who study other groups, this difference is perhaps found in the descriptive periphery, the particular subsets of descriptive content that is most relevant to their investigations, rather than the definitional core.

Of particular note here is the role and value of vagueness. In the case of species taxa, there is vagueness partly because species taxa have vague boundaries. It is not precisely clear at what point a new species lineage has appeared in the divergent change associated with evolution. Nor are the boundaries clear and distinct between the populations that constitute species and those that constitute subspecies or varieties. But there is also a vagueness due to the definitional structure of *species*. Even if the definitional core is set by evolutionary theory, there will always be different subsets of the descriptive periphery that may be relevant, depending on the difference among organisms and the divergent theoretical interests of different researchers. Furthermore, that descriptive periphery may change depending on new discoveries about things that are plausibly species lineages. Moreover, there may be a vagueness due to the failure of the theoretical framework to clearly specify reference by its definitional core. As it is not clear whether both light water and heavy water are truly *water*, according to modern chemical theory, and it is not clear whether nephrite and jadeite are truly *jade*, similarly, it is not clear what kinds of segments of lineages truly *count* as species on modern evolutionary theory.

The solution presented here can plausibly be regarded as a *species category essentialism*. The species category has an essence – set of necessary and sufficient definitional conditions – revolving around the idea of segments of population lineages. This is a qualified essentialism however, since the definitional core can always be revised in light of theoretical change or empirical discovery, and there is a vagueness in the definitional core. Species taxa, on the other hand, do not have such essences. As I argued in chapter 6, they should be regarded as *individuals* with organisms as parts, rather than as *sets* with organisms as members. This is the third thread of the solution, and has value insofar as it renders thinking about species more coherent with evolutionary theory. It is also useful heuristically – as a guide in how to extend and develop thinking about species, in terms of the processes that operate in the birth and death of species in speciation events, as well as the processes of change and cohesion that make them the fundamental units of evolution.

There are several reasons to take this proposed solution seriously. First, it is grounded on the actual usage of the terms *eidos* and the Latin translation *species*. There has been an implied division of conceptual labor in the thinking about species beginning with the early naturalists such as Ray and Linnaeus. Genealogical relations have been central in the understanding of the term *species* in its definitional core, even as other factors related to morphology and interbreeding have nonetheless been relevant. The second reason to take this solution seriously is that it explains the long endurance of the species problem. Since Darwin at least, we haven't been confused so much about what species are as with the functioning of the species concept relative to meaning and reference, and within the conceptual framework. This solution also explains why new empirical discoveries have not solved the species problem. It is a conceptual, not an empirical problem. Yet another reason to take this solution seriously is that it preserves the theoretical significance of species. If evolutionary theory tells us that species are the fundamental units of evolution, we should be worried if our solution to the species problem tells us otherwise. Finally, this solution gives us the conceptual resources to make sense out of other, related philosophical problems and puzzles. I have the space here to just mention them, but we can see where the framework developed here can profitably lead us.

HUMAN NATURE

One obvious area where the analysis here might be useful is in understanding particular species taxa. If the species category has an essence expressed by the definitional core, that species taxa are segments of population lineage – then we can look at individual taxa with this in mind. Most important to us is the species taxon *Homo sapiens*. What the framework developed here tells us is that we can use what we know about the relevant features of *species* population lineages to understand human nature. The most obvious insight is the failure of the essentialist approaches to species taxa. Humans don't have essences in either the classical or cluster sense. There is no set of necessary and sufficient *intrinsic* traits that make us members of the set *human*. Nor is there a disjunctive cluster set of traits. We should therefore quit looking for such human essences as the possession of reason, language or some kind of human dignity. And while we can think of human nature in terms of the historical and relational essences of historical kinds, this approach is less helpful in an inquiry into human nature than the approach that conceives of *Homo sapiens* as a spatio-temporally restricted individual. It merely tells us that humans are connected genealogically. The *species-as-individuals* thesis, on the other hand, tells us much more. It tells us to think about what makes *Homo sapiens* an *individual* – its origins, population structure, the processes governing change and the sources of cohesion.

Most notably, what the species-as-individuals thesis tells us is that there is a structure to the human population at any given time. There will be variability within any geographic area, as well as a typically greater variability across geographic areas. But humans are also similar at any given time, because of their history and perhaps stabilizing processes such as selection. Humans are even more highly variable over time. What might be "normal" relative to human nature, then, is not a particular set of traits, but particular patterns of variability. This emphasis on synchronic and diachronic variability is obvious to anyone who knows something about evolutionary theory and takes it seriously. There are other emphases suggested by the species-as-individuals thesis that are less obvious, but equally valuable in understanding human nature. One such emphasis is the cohesion of species *individuals*. This approach begs us to ask about the functioning of cohesive processes among individual humans. The answer to this question is far reaching, pointing beyond mere interbreeding to language, concept use, and the practice of science itself.

One obvious process of human cohesion is language use. Just as members of bird species recognize each other through song, members of the human species recognize each other through human language. Even if we do not all speak the same *human* language, nonetheless this is a cohesive factor because we recognize that we *could* communicate through language with any of our conspecifics. And even while we might understand the "languages" of other species, acquiring an interpretation of their songs, chirps or barks, we cannot speak those languages. The use of different human languages points to another aspect of *Homo sapiens* that we have already considered – its demic structure. Not all individuals in the human population lineage interact. They interact in smaller nested groupings – in demes. There are demes of language use – those who use the same language and communicate with each other. But there are also demes of other kinds, including those that function in science. This idea, in conjunction with other biological assumptions about selection and the genetic basis of development, is most forcefully advocated by David Hull:

> We are a social species. Our preference for living in herds is very likely to have some genetic basis. Since science is a social institution with its own norms and organization, our tendency to form groups also contributes to it. Even our general ability to use language must have some genetic basis, and language is yet another prerequisite for science. (Hull 2001: 97–98)

We can therefore understand science through the framework that conceives species as individuals, and asks us to look toward the processes of cohesion that make *Homo sapiens* an individual.

The basic idea here, the cohesion of humans into a species population with subpopulations or demes, is presupposed by my analysis here. We *could have* taken the various uses of the terms *eidos* and *species* to be independent and unconnected. But the historical and cohesive forces operating in the human population lineage suggest that the use of these terms is historically connected – the medieval thinkers read and discussed the views of Aristotle; Darwin read and thought about the theologically inclined naturalists; and modern systematists read and debated the views of those who preceded them. The terms are also connected within and among populations. Currently philosophers debate with each other about natural kinds and species, using their own patterns of discourse and vocabulary. Systematists likewise debate among themselves the meaning of the term *species* and competing views about species groupings, using their distinctive patterns of discourse and vocabulary. And at a different level, philosophers and systematists engage in debate with each other about these same issues, using slightly different patterns of discourse and vocabulary. The debates about the species problem can themselves be better understood by thinking about cohesive factors implicit in the species-as-individual thesis.

The analysis here might be useful in another way, by providing resources for the general analysis of scientific concepts. What can the details of the species debate tell us about the functioning of scientific concepts in general? By asking this question, we need not just assume that the factors useful here for understanding the species problem are also useful for understanding the functioning of other concepts in science. It may be that the details of the species problem are sufficiently different from the use of other concepts that no general conclusions can be drawn. On the other hand, the ideas developed here – the division of conceptual labor, definitional structure and reference potential – are all of potential use in understanding other cases of conceptual function and change. That value, if there is any, will play out in any attempts to employ these ideas.

A HISTORICAL METAPHYSICS

There is another potential value to the framework developed here. That is in its role within a general, historical metaphysical approach, an approach developed most philosophically by David Hull (Hull 1988, 1989, 2001). A long-dominant metaphysical framework, associated with Plato and his essentialism, is committed to the assumption that what is most real is that which is unchanging and timeless. This is what we saw in the essentialism that conceived of species taxa in terms of an unchanging set of essential conditions. The problem with that approach was that it conceived things that change and evolve within a framework that does not obviously accommodate change, or, at the very least, does not give a positive framework for understanding change. The metaphysical framework that tries to conceive the fundamental things that exist as natural kinds cannot easily accommodate things that change and that have a history. The idea of individuality used here can potentially provide a framework for thinking about the things in nature that change, within a framework that can naturally accommodate and explain change.

The framework adopted here can also provide models for thinking about change beyond biological evolution. Most relevant to the topic here, it can give us a model for thinking about the species concept itself. We can treat it as an *individual* with a history, a source of cohesion and change over time. The cohesion can plausibly be found in the definitional structure and the theoretical frameworks that give it structure. Similarly we can treat theories as spatio-temporal individuals, with historical trajectories and cohesion. And we can treat disciplines – even science itself – within this historical framework. Obviously we cannot address these topics in any detail here. But perhaps we can see the potential value of a metaphysical framework that is explicitly historical, and that inclines us to think about things that change in profitable ways. Perhaps our thinking about biological species can help us in thinking more profitably about a world that changes, and the things in it that change.

Bibliography

Agapow, Paul-Michael, O. R. P. Bininda-Edmunds, K. A. Crandall, J. L. Gittleman, G. M. Mace, J. C. Marshall, and A. Purvis (2004) "The Impact of Species Concept on Biodiversity Studies," *The Quarterly Review of Biology*, vol. 79, no. 2, 161–179.

Amundson, Ron, (1998) "Typology Reconsidered: Two Doctrines on the History of Evolutionary Biology," *Biology and Philosophy*, 13: 153–177.

Aristotle, (1995) *The Complete Works of Aristotle Vols. I and II*, J. Barnes, ed., Princeton University Press.

Ashworth, E. J. (2003) "Language and Logic," *Cambridge Companion to Medieval Philosophy*, A. S. McGrade, ed., Cambridge University Press, 73–96.

Balme, D. M. (1987a) "Aristotle's Biology was not Essentialistic," *Philosophical Issues in Aristotle's Biology*, A. Gotthelf and J. G. Lennox, eds., Cambridge University Press, 291–312.

(1987b) "Aristotle's Use of Division and Differentiae," *Philosophical Issues in Aristotle's Biology*, A. Gotthelf and J. G. Lennox, eds., Cambridge University Press, 69–89.

(1987c) "The Place of Biology in Aristotle's Philosophy," *Philosophical Issues in Aristotle's Biology*, A. Gotthelf and J. G. Lennox, eds., Cambridge University Press, 9–20.

Barker, Matthew J. (2007) "The Empirical Inadequacy of Species Cohesion by Gene Flow," *Philosophy of Science*, 74: 654–665.

Barnes, Jonathan (1995) "Life and Work," *The Cambridge Companion to Aristotle*, Cambridge University Press, 1–26.

Beatty, John (1992) "Speaking of Species: Darwin's Strategy," *The Units of Evolution: Essays on the Nature of Species*, M. Ereshefsky ed., Cambridge, MA: Bradford Books, 227–245.

Bird, Alexander (2000) *Thomas Kuhn*, Princeton University Press.

Bird, Alexander, and Emma Tobin (2008) "Natural Kinds," *The Stanford Encyclopedia of Philosophy* (Summer 2009 edn.), Edward N. Zalta ed., http://plato.stanford.edu/archives/sum2009/entries/natural-kinds/

Blackburn, Simon (2002) "Metaphysics," *The Blackwell Companion to Philosophy*, N Bunnin and E. P. Tsui-James, eds., Malden, MA: Blackwell Publishers, 61–89.

221

Bibliography

Blumenthal, Henry J. (1990) "Neoplatonic Elements in the *de Anima* Commentaries," *Aristotle Transformed*, R. Sorabji, ed., Ithaca, NY: Cornell University Press, 305–324.

(1996) *Aristotle and Neoplatonism in Late Antiquity: Interpretations of De Anima*, Ithaca, NY: Cornell University Press.

Boler, John F. (1963) "Abailard and the Problem of Universals," *Journal of the History of Philosophy* 1:104–126.

Bolton, Robert (1987) "Definition and Scientific Method in Aristotle's *Posterior Analytics* and *Generation of Animals*," *Philosophical Issues in Aristotle's Biology*, A. Gotthelf and J. G. Lennox, eds., Cambridge University Press, 120–166.

Boyd, Richard (1999) "Homeostasis, Species, and Higher Taxa," *Species: New Interdisciplinary Essays*, Robert A. Wilson, ed., Cambridge, MA: Massachusetts Institute of Technology Press, 141–185.

Brigandt, Ingo (2003) "Species Pluralism Does Not Imply Species Eliminativism," *Philosophy of Science*, 70: 1305–1316.

Broberg, Gunnar (1983) "Homo Sapiens: Linnaeus's Classification of Man," *Linnaeus the Man and His Work*, T. Frängsmyr, ed., Berkeley, CA: University of California Press, 156–194.

Brogaard, Berit (2004) "Species as Individuals," *Biology and Philosophy*, 19: 223–243.

Brown, Stephen F. (1999) "Realism versus Nominalism," *The Columbia History of Western Philosophy*, New York: Columbia University Press, 271–278.

Burian, Richard M. (1985) "On Conceptual Change in Biology: The Case of the Gene," *Evolution at a Crossroads: The New Biology and the New Philosophy of Science*," D. J. Depew and B. H. Weber, eds., Cambridge, MA: Massachusetts Institute of Technology Press, 21–42.

Burkhardt, Frederick and Sydney Smith, eds. (1987) *The Correspondence of Charles Darwin*, vol. II. Cambridge University Press.

(1990) *The Correspondence of Charles Darwin*, vol. VI. Cambridge University Press.

Cain, A. J. (1993) *Animal Species and Their Evolution*, Princeton University Press.

Carnap, Rudolf (1966) *An Introduction to the Philosophy of Science*, M. Gardner, ed., New York: Basic Books.

Claridge, M. F., H. A. Dawah and M. R. Wilson (1997) "Practical Approaches to Species Concepts for Living Organisms," *Species: The Units of Biodiversity*, M. F. Claridge, H. A. Dawah, M. R. Wilson, eds., London: Chapman and Hall, 1–15.

Cracraft, Joel (1992) "Species Concepts and Speciation Analysis," *The Units of Evolution: Essays on the Nature of Species*, M. Ereshefsky ed., Cambridge, MA: Bradford Books, 93–120.

(1997) "Species Concepts in Systematics and Conservation Biology – an Ornithological Viewpoint," *Species: The Units of Biodiversity*, M. F. Claridge, H.A. Dawah, M. R. Wilson, eds., London: Chapman and Hall, 325–339.

(2000) "Species Concepts in Theoretical and Applied Biology: A Systematic Debate with Consequences," *Species Concepts and Phylogenetic Theory*,

Q. D. Wheeler, Quentin and R. Meier, eds., New York: Columbia University Press, 3–14.

Darwin, Charles (1851) *The Complete Works of Charles Darwin Online*, University of Cambridge, http://darwin-online.org.uk/content/frameset?vi ewtype=text&itemID=F339.1&keywords=cirripedia&pageseq=1.

(1964) *On the Origin of Species: A Facsimile of the First Edition*, Cambridge, MA: Harvard University Press.

de Queiroz, Kevin (1999) "The General Lineage Concept of Species and the Defining Properties of the Species Category," *Species: New Interdisciplinary Essays*, Robert A. Wilson, ed., Cambridge, MA: Massachusetts Institute of Technology Press, 49–89.

(2005) "Ernst Mayr and the Modern Concept of Species," *Proceedings of the National Academy of Sciences*, vol. CII, suppl. 1, 6600–6607.

de Vries, Hugo (1905) *Species and Varieties, Their Origin and Mutation*, Chicago, IL: Open Court Publishing.

Dennett, Daniel (1995) *Darwin's Dangerous Idea*, New York: Simon and Schuster.

Devitt, Michael (2008) "Resurrecting Biological Essentialism," *Philosophy of Science*, vol. 75, no. 3, 344–382.

Dobzhansky, Theodosius (1937) *Genetics and the Origin of Species*, New York: Columbia University Press.

Dupré, John (1993) *The Disorder of Things: Philosophical Foundations for the Disunity of Science*, Cambridge, MA: Harvard University Press.

(1999) "On the Impossibility of a Monistic Account of Species," *Species: New Interdisciplinary Essays*, Robert A. Wilson, ed., Cambridge, MA: Massachusetts Institute of Technology Press, 3–19.

Ebbesen, Stan (1990a) "Boethius as an Aristotelian Commentator," *Aristotle Transformed*, R. Sorabji, ed., Ithaca, NY: Cornell University Press, 373–391.

(1990b) "Philoponus, Alexander and the Origins of Medieval Logic," *Aristotle Transformed*, R. Sorabji, ed., Ithaca, NY: Cornell University Press, 445–461.

(1990c) "Porphyry's Legacy to Logic: A Reconstruction," *Aristotle Transformed*, R. Sorabji, ed., Ithaca, NY: Cornell University Press, 141–179.

Eldredge, Niles (1995) "Species, Selection, and Paterson's Concept of the Specific-Mate Recognition System," *Speciation and the Recognition Concept*, Baltimore, MD: The Johns Hopkins University Press, 464–477.

Eldredge, Niles, and Joel Cracraft (1980) *Phylogenetic Patterns and the Evolutionary Process*, New York: Columbia University Press.

Ereshefsky, Marc (1992) "Species, Higher Taxa, and the Units of Evolution," *The Units of Evolution: Essays on the Nature of Species*, M. Ereshefsky, ed., Cambridge, MA: Bradford Books, 381–398.

(2001) *The Poverty of the Linnaean Hierarchy: A Philosophical Study of biological Taxonomy*, Cambridge University Press.

Eriksson, Gunnar (1983) "Linnaeus the Botanist," *Linnaeus the Man and His Work*, T. Frängsmyr, ed., Berkeley, CA: University of California Press, 63–109.

Bibliography

Falcon, Andrea (2005) "Commentators on Aristotle," *The Stanford Encyclopedia of Philosophy* (Winter 2005 edn), Edward N. Zalta ed., http://plato.stanford.edu/archives/fall2005/entries/aristotle-commentators/.

Frängsmyr, Tore, ed. (1983) *Linnaeus, the Man and His Work,* Berkeley, CA: University of California Press.

Frede, Dorothea (2009) "Alexander of Aphrodisias," *The Stanford Encyclopedia of Philosophy* (Summer 2009 edn), Edward N. Zalta ed., http://plato.stanford.edu/archives/sum2009/entries/alexander-aphrodisias/.

Frege, Gottlob (1990) "On Sense and Nominatum," reprinted in *The Philosophy of Language*, A.P. Martinich, ed., Oxford University Press, 190–202.

Ghiselin, Michael, T. (1969) *The Triumph of the Darwinian Method*, University of Chicago Press.

(1989) "Individuality, History and the Laws of Nature in Biology," *What the Philosophy of Biology Is: Essays Dedicated to David Hull*, M. Ruse, ed., Dordrect, The Netherlands: Kluwer Academic Publishers.

(1997) *Metaphysics and the Origin of Species*, Albany, NY: State University of New York Press.

Gopnik, Alison, and Andrew N. Meltzoff (1997) *Words, Thoughts, and Theories*, Cambridge, MA: Massachusetts Institute of Technology Press.

Gotthelf, Allan (1987) "First Principles in Aristotle's *Parts of Animals*," *Philosophical Issues in Aristotle's Biology*, A. Gotthelf and J. G. Lennox, eds., Cambridge University Press, 167–198.

Grene, Marjorie, and David Depew (2004) *The Philosophy of Biology: An Episodic History*, Cambridge University Press.

Griffiths, Paul (1999) "Squaring the Circle, Natural Kinds with Historical Essences," *Species, New Interdisciplinary Essays*, Robert A. Wilson, ed., Cambridge, MA: Massachusetts Institute of Technology Press, 209–228.

Hadot, Pierre (1990) "The Harmony of Plotinus and Aristotle According to Porphyry," *Aristotle Transformed*, R. Sorabji, ed., Ithaca NY: Cornell University Press, 125–140.

Hankinson, R. J. (1994) "Galen and the Logic of Relations," *Aristotle in Late Antiquity*, L. Schrenk, ed., Washington, DC: The Catholic University of America Press, 57–75.

Hennig, Willi (1977) *Phylogenetic Systematics*, Urbana, IL: University of Illinois Press.

Hey, Jody (2001) *Genes, Categories and Species: the Evolutionary and Cognitive Causes of the Species Problem*, Oxford University Press.

Hull, David L. (1983) *Darwin and His Critics*, University of Chicago Press.

(1988) *Science as Process: an Evolutionary Account of the Social and Conceptual Development of Science*, University of Chicago Press.

(1989) *The Metaphysics of Evolution*, Albany, NY: State University of New York Press.

(1992a) "The Effect of Essentialism on Taxonomy: Two Thousand Years of Stasis," *The Units of Evolution: Essays on the Nature of Species*, M. Ereshefsky, ed., Cambridge, MA: Bradford Books, 199–225.

Bibliography

(1992b) "A Matter of Individuality," *The Units of Evolution: Essays on the Nature of Species*, M. Ereshefsky, ed., Cambridge, MA: Bradford Books, 293–316.

(1997). "The Ideal Species Concept – and Why We Can't Get It," in *Species: the Units of Biodiversity*, M. F. Claridge, H. A. Dawah, M. R. Wilson, eds., London: Chapman and Hall, 357–380.

(1999) "On the Plurality of Species: Questioning the Party Line," *Species: New Interdisciplinary Essays*, Robert A. Wilson, ed., Cambridge, MA: Massachusetts Institute of Technology Press, 23–48.

(2001) *Science and Selection: Essays on Biological Evolution and the Philosophy of Science*, Cambridge University Press.

Huxley, Julian (1940) *The New Systematics*, London Oxford University Press.

Hyman, Arthur and James J. Walsh (1983) *Philosophy in the Middle Ages*, Indianapolis, IN: Hackett Publishing Company.

Jones, W. T. (1969) *The Medieval Mind*, San Diego, CA.: Harcourt Brace Jovanovich.

Kitcher, Philip (1978) "Theories, Theorists and Theoretical Change," *Philosophical Review*, 87: 519–547.

(1984) "Against the Monism of the Moment," *Philosophy of Science*, vol. 51, no. 4. (December 8), 616–630.

(1992) "Species," *The Units of Evolution: Essays on the Nature of Species*, M. Ereshefsky, ed., Cambridge, MA: Bradford Books, 317–341.

(1998) "Explanatory Unification," *Introductory Readings in the Philosophy of Science*, E. D. Klemke, R. Hollinger, D. W. Rudge, eds., Amherst, NY: Prometheus Books, 278–301.

Kitts, David R. and David J. Kitts (1979) "Biological Species as Natural Kinds," *Philosophy of Science*, vol. 46, no. 4, 613–622.

Klima, Gyula (2003) "Natures: The Problem of Universals," *Cambridge Companion to Medieval Philosophy*, A. S. McGrade, ed., Cambridge University Press, 196–207.

Kripke, Saul (1972) *Naming and Necessity*, Cambridge, MA: Harvard University Press.

Kuhn, Thomas (1977) *The Essential Tension*, The University of Chicago Press.

(1996) *The Structure of Scientific Revolutions*, 3rd edn,The University of Chicago Press.

LaPorte, Joseph, (2004) *Natural Kinds and Conceptual Change*, Cambridge University Press.

Larson, James L. (1968) "The Species Concept of Linnaeus," *Isis*, vol. 59, no. 3, 291–299.

(1971) *Reason and Experience: The Representation of Natural Order in the Work of Carl von Linne*, Berkeley, CA: University of California Press.

Lauden, Larry (1984) *Science and Values: The Aims of Science and Their Role in Scientific Debate*, Berkeley, CA: University of California Press.

Lennox, James (1980) "Aristotle on Genera, Species and the More and the Less," *Journal of the History of Biology*, 13, 321–346.

Bibliography

(1987a) "Divide and Explain: The Posterior Analytics in Practice," *Philosophical Issues in Aristotle's Biology*, A. Gotthelf and J. G. Lennox, eds., Cambridge University Press, 90–119.

(1987b) "Kinds, Forms of Kinds and the More and Less in Aristotle's Biology," *Philosophical Issues in Aristotle's Biology*, Gotthelf and Lennox, eds., Cambridge University Press, 339–359.

(2006) "Aristotle's Biology," *The Stanford Encyclopedia of Philosophy (Fall 2006 edn)*, Edward N. Zalta, ed., http://plato.stanford.edu/archives/fall2006/entries/aristotle-biology/.

Lewis, David (1983), "Extrinsic Properties," *Philosophical Studies*, 44: 197–200.

Lindroth, Sten (1983) "The Two Faces of Linnaeus," *Linnaeus the Man and His Work*, T. Frängsmyr, ed., Berkeley, CA: University of California Press, 1–62.

Linnaeus, Carolus (1964) *Systema Naturae, Facsimile of the First Edition*, M. J. S. Engel-Ledeboer and H. Engel, eds., Nieuwkoop, Netherlands: B. De Graaf. Publ.

Lovejoy, Arthur O. (1968) "Buffon and the Problem of Species," *Forerunners of Darwin*, Baltimore, MD: Johns Hopkins University Press, 84–113.

Lyell, Charles (1997) *Principles of Geology*, London Penguin Books.

MacLaurin, James and Kim Sterelny (2008) *What is Biodiversity?*, University of Chicago Press.

Madigan, Arthur S. J. (1994) "Alexander on Aristotle's Species and Genera as Principles," *Aristotle in Late Antiquity*, L. Schrenk, ed., Washington DC: The Catholic University of America Press, 76–91.

Magnello, M. Eileen (1996) "Karl Pearson's Gresham Lectures: W. F. R. Weldon, Speciation and the Origin of Pearsonian Statistics," *British Journal for the History of Science*, vol. 29, no. 1 (March), 43–63.

Mallet, James (1995) "A Species Definition for the Modern Synthesis," *Trends in Ecology and Evolution*, vol. 10, no. 7, 294–299.

Marenbon, John (2005) "Anicius Manlius Severinus Boethius," *The Stanford Encyclopedia of Philosophy* (Fall 2008 edn), Edward N. Zalta, ed., http://plato.stanford.edu/archives/fall2008/entries/boethius/.

Margolis, Eric and Stephen Laurence (2006) "Concepts," *The Stanford Encyclopedia of Philosophy* (Spring 2007 edn), Edward N. Zalta, ed., http://plato.stanford.edu/archives/spr2007/entries/concepts/.

Marrone, Steven P. (2003) "Medieval Philosophy in Context," *Cambridge Companion to Medieval Philosophy*, A. S. McGrade, ed., Cambridge University Press, 10–50.

Mayden, Richard L. (1997) "A Hierarchy of Species Concepts: The Denouement in the Saga of the Species Problem," *Species: The Units of Biodiversity*, M. F. Claridge, H. A. Dawah, M. R. Wilson, eds., London: Chapman and Hall, 381–424.

(1999) "Consilience and a Hierarchy of Species Concepts: Advances Toward Closure on the Species Puzzle," *Journal of Nematology*, vol. 31, no. 2, 95–116.

Mayr, Ernst (1942) *Systematics and the Origin of Species*, New York: Columbia University Press.

(1957a) "Difficulties and Importance of the Biological Species," *The Species Problem*, Mayr, ed., Washington, DC: American Association for the Advancement of Science, 371–388.

(1957b) "Species Concepts and Definitions," *The Species Problem*, Mayr, ed., Washington, DC: American Association for the Advancement of Science, 1–22.

(1982) *The Growth of Biological Thought*, Cambridge, MA, Belknap Press.

(1992) "Species Concepts and Their Application," *The Units of Evolution: Essays on the Nature of Species*, M. Ereshefsky, ed., Cambridge, MA: Bradford Books, 15–25.

(2000) "The Biological Species Concept," *Species Concepts and Phylogenetic Theory*, Q. D. Wheeler, Quentin and R. Meier, eds., New York: Columbia University Press, 17–29.

Mayr, Ernst and Peter D. Ashlock (1991) *Principles of Systematic Zoology*, New York: McGraw-Hill.

Mellor, D. H. (1977) "Natural Kinds," *The British Journal for the Philosophy of Science*, vol. 28, no. 4, 299–312.

Mellor, D. H. and Alex Oliver (1997) *Properties*, Oxford University Press.

Mishler, Brent D. and Robert Brandon (1987) "Individuality, Pluralism, and the Phylogenetic Species Concept", *Biology and Philosophy*, 2: 397–414.

Mishler, Brent D. and Michael Donoghue (1992) "Species Concepts: A Case for Pluralism," *The Units of Evolution: Essays on the Nature of Species*, M. Ereshefsky, ed., Cambridge, MA: Bradford Books, 121–137.

Mishler, Brent D., and Edward C. Theriot (2000) "The Phylogenetic Species Concept (*sensu* Mishler and Theriot): Monophyly, Apomorphy and the Phylogenetic Species Concept," *Species Concepts and Phylogenetic Theory*, Q. D. Wheeler, Quentin and R. Meier, eds., New York: Columbia University Press, 44–54.

Mueller, Ian "Aristotle's Doctrine of Abstraction in the Commentators," *Aristotle Transformed*, R. Sorabji, ed., Ithaca NY: Cornell University Press, 463–480.

Murphy, Gregory L., and Douglas L. Medin (1985) "The Role of Theories in Conceptual Coherence," *Psychological Review*, vol. 92, no. 3 (July 1985), 289–316.

National Research Council Committee on Scientific Issues in the Endangered Species Act (1995) *Science and the Endangered Species Act*, Washington, DC: National Academies Press.

Newton-Smith, W. H. (1994) *The Rationality of Science*, London: Routledge Press.

Niklas, Karl J. (1997) *The Evolutionary Biology of Plants*, University of Chicago Press.

Ogilvie, Brian W. (2006) *The Science of Describing: Natural History in Renaissance Europe*, University of Chicago Press.

Panchen, Alec L. (1992) *Classification, Evolution, and the Nature of Biology*, Cambridge University Press.

227

Bibliography

Paterson, Hugh (1992) "The Recognition Concepts of Species," *The Units of Evolution: Essays on the Nature of Species*, M. Ereshefsky, ed., Cambridge, MA: Bradford Books, 139–158.

(1993) *Evolution and the Recognition Concept of Species, Collected Writings of H. E. H. Paterson*, S. F. McEvey, ed., Baltimore, MD: Johns Hopkins University Press.

Peacocke, Christopher (1992) *A Study of Concepts*, Cambridge, MA: Massachusetts Institute of Technology Press.

Pellegrin, Pierre (1987) "Logical Difference and Biological Difference: The Unity of Aristotle's Thought," *Philosophical Issues in Aristotle's Biology*, A. Gotthelf and J. G. Lennox, eds., Cambridge University Press, 313–338.

Peters, F. E. (1967) *Greek Philosophical Terms: A Historical Lexicon*, New York University Press.

Purvis, O. W. (1997) "The Species Concept in Lichens," *Species: The Units of Biodiversity*, M. F. Claridge, H.A. Dawah, M. R. Wilson, eds., London: Chapman and Hall, 109–134.

Putnam, Hilary (1990) "Meaning and Reference," reprinted in *The Philosophy of Language*, 2nd edn, A. Martinich, ed., New York: Oxford University Press, 308–315.

Quine, W. V. O. (1976) *The Ways of Paradox and Other Essays*, Cambridge, MA: Harvard University Press.

(1990) "Two Dogmas of Empiricism," reprinted in *The Philosophy of Language*, 2nd edn, A. Martinich, ed., New York: Oxford University Press, 26–39.

Ray, John (1735) *The Wisdom of God Manifested in the Works of Creation*, 10th edn, London: William Innys and Richard Manby.

Reydon, Thomas A. C. (2003) "Discussion: Species Are Individuals – Or Are They?", *Philosophy of Science*, 70: 49–56.

Richards, Janet Radcliffe (2000) *Human Nature after Darwin*, London: Routledge.

Richards, Richard A. (2002) "Kuhnian Values and Cladistic Parsimony," *Perspectives on Science*, vol. 10, no. 1, 1–27.

(2003) "Character Individuation in Phylogenetic Inference," *Philosophy of Science*, 70, 264–279.

(2005) "Evolutionary Naturalism and the Logical Structure of Valuation: The Other Side of Error Theory," *Cosmos and History: The Journal of Natural and Social*, vol. 1, no. 2, 270–294.

(2007a) "Solving the Species Problem: Kitcher and Hull on Sets and Individuals," *Philosophy of Biology*, Michael Ruse. ed., Amherst, NY: Prometheus Books.

(2007b) "The Species Problem: A Philosophical Analysis," *The Encyclopedia of Life Sciences*, Chichester: John Wiley.

(2008) "Species and Taxonomy," *Oxford Handbook of the Philosophy of Biology*, Oxford University Press.

Ruse, Michael (1987) "Biological Species: Natural Kinds, Individuals or What?" *The British Journal for the Philosophy of Science*, vol. 38, no. 2, 225–242.

228

Bibliography

(1992) "Biological Species: Natural Kinds, Individuals, or What?", *The Units of Evolution: Essays on the Nature of Species*, M. Ereshefsky, ed., Cambridge, MA: Bradford Books, 343–361.

(1999) *The Darwinian Revolution*, University of Chicago Press.

(1998) *Taking Darwin Seriously: A Naturalistic Approach to Philosophy*, Amherst, NY: Prometheus Books.

Sharples, Robert W. (1990) "The School of Alexander?" *Aristotle Transformed*, R. Sorabji, ed., Ithaca, NY: Cornell University Press, 83–111.

Shiel, James (1990) "Boethius' Commentaries on Aristotle," *Aristotle Transformed*, R. Sorabji, ed., Ithaca, NY: Cornell University Press, 349–372.

Simpson, George Gaylord (1951) "The Species Concept," *Evolution*, vol. 5, no. 4, 285–298.

(1961) *Principles of Animal Taxonomy*, New York: Columbia University Press.

Sloan, Philip R. (1976) "The Buffon-Linnaeus Controversy," *Isis*, vol. 67, no. 3, 356–375.

(1979), "Buffon, German Biology, and the Historical Interpretation of Biological Species," *The British Journal for the History of Science*, vol. 12, no. 2, 109–153.

(1985) "From Logical Universals to Historical Individuals," *Histoire du Concept d'Espèce dans les Sciences de la Vie*, Paris: Fondation Singer-Polignac.

(2003) "Reflections on the Species Problem," *The Philosophy of Marjorie Grene*, R. Auxier and L. E. Hahn, eds., Chicago, IL.: Open Court.

Slobodchikoff, C. N. (1976) *Concepts of Species*, Stroudsberg, PA: Dowden, Hutchinson and Ross.

Sneath, P. H. A., and Robert R. Sokal (1973) *Numerical Taxonomy*, San Francisco, CA: W. H. Freeman.

Soames, Scott (2005) *Reference and Description: The Case against Two-Dimensionalism*, Princeton University Press.

Sober, Elliott (1980) "Evolution, Population Thinking, and Essentialism," *Philosophy of Science*, vol. 47, no. 3, 350–383.

(1984) "Sets, Species and Evolution: Comments on Philip Kitcher's 'Species'," *Philosophy of Science*, vol. 51, no. 2, 334–341.

(1992) "Evolution, Population Thinking, and Essentialism," *The Units of Evolution: Essays on the Nature of Species*, M. Ereshefsky ed., Cambridge, MA: Bradford Books, 247–278.

(2000) *Philosophy of Biology, 2nd edn*, Boulder, CO: Westview Press.

Sorabji, Richard (1990) "The Ancient Commentators on Aristotle," *Aristotle Transformed*, R. Sorabji, ed., Ithaca, NY: Cornell University Press, 1–30.

Spade, Paul Vincent (2008) "William of Ockham," *The Stanford Encyclopedia of Philosophy* (Fall 2008edn), Edward N. Zalta ed., http://plato.stanford.edu/archives/fall2008/entries/ockham/.

Stafleu, Frans A. (1971) *Linnaeus and the Linnaeans*, Utrecht, Netherlands: A. Oosthooek's Uitgeversmaatschappij.

Stamos, David, (2003) *The Species Problem*, Lanhan, MD: Lexington Books.

(2007) *Darwin and the Nature of Species*, Albany, NY: SUNY Press.

Bibliography

Stauffer, Robert C. (1975) *Charles Darwin's Natural Selection*, Cambridge University Press.

Stott, Rebecca (2003) *Darwin and the Barnacle*, New York: W. W. Norton and Company.

Templeton, Alan (1992) "The Meaning of Species and Speciation: A Genetic Perspective," *The Units of Evolution: Essays on the Nature of Species*, M. Ereshefsky, ed., Cambridge, MA: Bradford Books, 159–183.

Thompson, Augustine (1995) "The Debate on Universals Before Peter Abelard," *The Journal of the History of Philosophy*, vol. 33, no. 3, 409–429.

Van Valen, Leigh (1992) "Ecological Species, Multispecies, and Oaks," *The Units of Evolution: Essays on the Nature of Species*, M. Ereshefsky, ed., Cambridge, MA: Bradford Books, 69–77.

Wheeler, Quentin D. and Norman I. Platnick (2000) "The Phylogenetic Species Concept (*sensu* Wheeler and Platnick)," *Species Concepts and Phylogenetic Theory*, Q. D. Wheeler, Quentin and R. Meier eds., New York: Columbia University Press, 55–69.

Whewell, William (1984) *Selected Writings on the History of Science*, University of Chicago Press.

(1989) *Theory of Scientific Method*, R. E. Butts, ed., Indianapolis, IN.: Hackett Publishing.

Wiley, E. O. (1981) *Phylogenetics: The Theory and Practice of Phylogenetic Systematics*, New York: John Wiley and Sons.

Wiley, E. O., and Richard L. Mayden (2000) "The Evolutionary Species Concept," *Species Concepts and Phylogenetic Theory*, Q. D. Wheeler, Quentin and R. Meier, eds., New York: Columbia University Press, 70–89.

Williams, Mary B. (1989) "Evolvers are Individuals: Extension of the Species as Individuals Claim," *What the Philosophy of Biology Is: Essays dedicated to David Hull*, M. Ruse, ed., Dordrect, The Netherlands: Kluwer Academic Publishers, 301–308.

Wilson, Edward O. (2000) *Sociobiology: The New Synthesis*, Cambridge, MA: Harvard University Press.

Wilson, Leonard G. (1970) *Sir Charles Lyell's Scientific Journals on the Species Question*, New Haven, CT: Yale University Press.

Wilson, Robert A. (1999) "Realism, Essence, and Kind: Resuscitating Species Essentialism?" *Species: New Interdisciplinary Essays*, Robert A. Wilson, ed., Cambridge, MA: Massachusetts Institute of Technology Press, 187–207.

Winsor, Mary (2003) "Non-essentialist Methods in Pre-Darwinian Taxonomy," *Biology and Philosophy*, 18: 387–400.

(2006a) "The Creation of the Essentialism Story: An Exercise in Metahistory," *History and Philosophy of the Life Sciences*, 28, 149–174.

(2006b) "Linnaeus's Biology was not Essentialist," *Annuals of the Missouri Botanical Garden*, 93: 2–7.

Wittgenstein, Ludwig (1968) *Philosophical Investigations*, New York: MacMillan Publishing Co.

Zimmer, Carl (2008) "What Is a Species," *Scientific American* (June), 72–79.

Zirkle, Conway (1959) "Species Before Dawin," *Proceedings of the American Philosophical Society*, vol. 103, no. 5, 636–644.

Index

231

Printed in the United States
By Bookmasters